Studies in Chemistry No. 6

Editors: Bryan J. Stokes and
Anthony J. Malpas

THE CHEMISTRY
OF FIBRES

J. E. McINTYRE, B.Sc

Section Manager
Research Department
ICI Fibres Limited

Edward Arnold

© J. E. McIntyre, 1971
First published 1971
by Edward Arnold (Publishers) Limited,
41 Maddox Street,
London, W1R 0AN

Boards edition ISBN: 0 7131 2327 3
Paper edition   ISBN: 0 7131 2328 1

Printed in Great Britain by
Richard Clay (The Chaucer Press), Ltd.,
Bungay, Suffolk

D
677.
MACI

Editors' Preface

Chemistry courses for schools, technical colleges and universities are sometimes criticized on the grounds that they disregard the applied aspects of the subject, and new syllabuses tend to give increasing emphasis on academic principles to the exclusion of industrial practice.

It is, of course, wholly desirable that special attention should be given to the basic principles of the subject, but the influence of chemistry upon industry, medicine, agriculture and indeed the whole structure of society is both immense and increasing. No chemical education should be thought complete without some informed acquaintance with the impact of chemistry on society, and, if the benefits of scientific progress are to continue, capable students must be stimulated to take up careers in applied fields.

One of the aims of the 'Studies in Chemistry' series of books is to provide background reading for students on the newer and more rapidly developing areas of applied chemistry. It is hoped that titles having this aim, of which this is one, will show the reader how chemical principles are applied to human needs, and give some idea of the nature and type of work which this operation involves.

1971   B. J. S.
                                                                                        A. J. M.

Preface

The industrial importance of fibres is obvious. Over twenty million tons of fibre are produced each year. Already half is man-made, over a quarter entirely synthetic. Even the production of chemical intermediates for synthetic fibres is a substantial chemical industry on its own: five million tons a year.

Fibre chemistry has some unusual attractions. In all polymers the relation between chemical structure and physical properties is particularly important. In fibres, orientation of polymer chains adds fascinating directional effects. Their study will lead to new fibres. It may even lead us full circle, to understand the growth of natural fibres, and through them of the many oriented structures within and around us.

1971   J. E. McI.

Contents

Editors' Preface   iii

Preface   iii

1 Historical introduction   1

2 Principles of fibre structure   3
    2.1   Fibres are oriented polymers   3
    2.2   Linear and branched polymers   6
    2.3   Chemical structure and fibre properties   6

3 Principles of fibre-making processes   12
    3.1   Spinning processes   12
    3.2   Filament yarn and staple   17

4 Fibres from natural polymers   19
    4.1   Natural cellulosic fibres   19
    4.2   Man-made fibres derived from cellulose   23
    4.3   Protein fibres   31

5 Fibres from synthetic polymers   38
    5.1   Fibre manufacture and properties   38
    5.2   Intermediates for synthetic fibres   50

6 Colour and lustre   58

7 The future   64

Some well-known trade-names   67

Further reading   68

1 Historical introduction

All the important natural fibres were established in textile use thousands of years ago: the bast fibres, such as flax, in Egypt at least 7000 years ago; silk in China, wool in Europe and the Middle East, cotton in India and the Americas at least 5000 years ago. Archaeologists have found evidence that even Stone Age man knew how to twist short lengths of fibre together to form cords and yarns, using a spinning process not very different from that used in primitive parts of the world to this day. Other textile processes, too, such as sewing, carding, weaving, bleaching and dyeing, predate written history.

No doubt process improvements were introduced throughout historical times, but progress was slow until the late eighteenth and early nineteenth centuries when mechanical improvements such as the spinning jenny (1764), Crompton's mule (1779), and the ring frame (1828), together with chemical improvements such as the use of bleaching powder (1799) transformed the textile industry from a domestic into a factory occupation, and brought cotton to the position it still holds as the most used of all textile fibres. But there were no new fibres, and little was known of the chemical structures of the existing fibres.

Two vital steps were needed before introduction of man-made fibres at the end of the nineteenth century—identification of a suitable raw material, and design of suitable equipment. Some of the stages leading to the first commercial man-made fibres, all forms of cellulose regenerated from soluble derivatives, are listed below:

1713 RÉAUMUR (France) draws fibres from the surface of molten glass, winds them on to a cylinder, and produces fabrics. Neither material nor equipment is suitable for development.

1832 Discovery of cellulose nitrate, a soluble cellulose derivative.

1840 Introduction of mechanical wood pulp, a cheap source of cellulose, for making paper.

1842 SCHWABE (U.K.) spins molten glass through fine nozzles, but fails to find a more suitable material.

1855 AUDEMARS (Switzerland) pulls fibres long enough to wind up from a mixed solution of cellulose nitrate and raw rubber, but does not try spinning through holes.

1857 SCHWEIZER (Switzerland) finds that cellulose can be dissolved in cuprammonium hydroxide solution.

1860 Introduction of chemical wood pulp for paper.

1862 OZANAM (U.K.) spins regenerated silk fibres by forcing silk solution through fine holes.

1880 SWAN (U.K.) extrudes cellulose nitrate in a controlled fashion through holes to produce electric light filaments.

1884–9 CHARDONNET (France) shows how to spin textile filaments from cellulose nitrate and regenerate cellulose by reducing the nitrate groups. He sets up commercial production of the first man-made fibre.

1891 CROSS, BEVAN, and BEADLE (U.K.) discover that cellulose is dissolved if treated with sodium hydroxide then carbon disulphide—the basis of viscose rayon.

1899 Commercial production in Germany of cuprammonium rayon, based on SCHWEIZER'S discovery of 1857.

1903 Commercial production in U.K. of viscose rayon, based on CROSS, BEVAN, and BEADLE'S discovery of 1891.

So far the new fibres consisted essentially of regenerated cellulose. Next to be developed were the acetic esters of cellulose. A little cellulose triacetate fibre was produced in the U.S.A. from about 1914–1927. Far more successful was the introduction in 1919–1921 by the Dreyfus brothers of fibres from secondary cellulose acetate, which became established as 'Celanese'.

The remarkable feature of fibre development up to this point is that most of the ideas about chemical structure of fibres were thoroughly inaccurate. Not until 1920 did Hermann Staudinger begin the work on the structure of giant molecules for which he was eventually awarded the Nobel Prize for Chemistry in 1953. When he started it was known that some colloidal materials appeared to have molecular weights of many thousands, but most chemists believed that such colloids were aggregates of small molecules held together by secondary forces of some kind. Staudinger's achievement was to prove that they really did consist of molecules of very high molecular weight. Once this concept was accepted and understood, the synthesis of new materials of this type, known as *polymers*, ceased to be a matter of chance and became a science.

One of the greatest practical exponents of the new ideas was W. H. Carothers, of the American chemical company E. I. du Pont de Nemours. His team found in 1932 that if certain synthetic polymers of sufficiently high molecular weight were melted, it was possible to pull fibres from the surface. Many polymers were investigated before useful products were located in the class known as polyamides. The first synthetic fibre, nylon-6,6, was launched in the U.S.A. in 1938. It was quickly followed by others: Perlon L (nylon-6) in Germany, and Vinyon, a fibre made from vinyl chloride and vinyl acetate polymerized together, in the U.S.A. These have since been joined by many others, particularly important being the aromatic polyesters and acrylics introduced in the late 1940's.

2 Principles of fibre structure

2.1 Fibres are oriented polymers

What is the difference between a fibre and any other form of matter? One answer is obvious: a fibre is very much longer in one direction than in the others. The cross-section is not necessarily round—silk, for example, is roughly triangular in cross-section—but whatever the shape the ratio of maximum to minimum diameter is low compared with the ratio of length to maximum diameter.

These are purely dimensional characteristics unrelated to chemical structure. The most important chemical feature is that fibres consist of polymers—compounds consisting of very large molecules that contain recurring structural units repeated many times within each molecule. For example, pure nylon-6 has the structure

$$H_2N(CH_2)_5CO(NH(CH_2)_5CO)_nNH(CH_2)_5COOH$$

$$\text{end group} \qquad \text{repeat unit} \qquad \text{end group}$$

The recurring structural unit, or *repeat unit*, is —$NH(CH_2)_5CO$—. A single molecule might contain a hundred such units ($n = 100$), making its molecular weight over 11 000. Such large molecules are known as *macromolecules*.

Most ordinary chemical compounds have a single, well-defined, low molecular weight; for example, glucose, $C_6H_{12}O_6$, has molecular weight 180. In polymers, on the other hand, the molecular weight is not only high but also different from molecule to molecule. The molecular weight of the polymer is an average of the molecular weights of the macromolecules present. This average can be calculated and measured in different ways, giving different numerical values, so it is important to state which average is being quoted. The two most important average molecular weights are known as weight average molecular weight ($\overline{M}_w$), and number average molecular weight ($\overline{M}_n$). $\overline{M}_w$ cannot be smaller than $\overline{M}_n$; it is usually $1\frac{1}{2}$–2 times $\overline{M}_n$, but can be very much greater.

Measurement of molecular weight is much more difficult than for simple compounds, and different techniques give different estimates. Osmotic pressure measurement on solutions and determination of end-groups both give estimates of $\overline{M}_n$; light scattering of solutions gives $\overline{M}_w$; and there is a comparatively recent technique for rapid fractionation known as gel permeation chromatography that gives a picture of the entire distribution of molecular weights. In everyday industrial practice, however, simple measurements of the viscosities of either dissolved or molten polymers are used to control the molecular weight. Calibration against one of the other techniques is necessary if the actual molecular weight of a sample must be known.

Merely to have a polymer is not to have a fibre. Moulded plastics are made from polymers but are clearly not fibres. Nor is it enough to make the physical dimensions conform to those typical of fibres. The important property that distinguishes polymers in fibre form is their orientation, or *anisotropy*. Isotropic materials have the same properties in all directions; anisotropic materials have different properties in different directions, and fibres in particular are much stronger lengthwise than transversely. The reason is that the individual molecules in the fibre are oriented. Consider a piece of string as a very rough model of a polymer molecule. If allowed to roll about in a drawer or a pocket, it will adopt a curled-up shape. Similarly a polymer molecule, if affected only by purely random forces, will curl up. But if the string is held extended with the two ends far apart, most of it will lie in a particular direction; if the ends are as far apart as possible, all the string will lie in one direction. In these two cases the string is oriented, partly in the first case, fully in the second case. Similarly, the chains of the polymer molecules in a fibre are extended along its lengthwise axis. The more extended they are, the more fully oriented they are said to be. If we now orient several pieces of string in the same direction and place them side by side, it is quite easy by pulling to abstract one or many of them. Clearly some force is present in the fibre that prevents the oriented chains from sliding over each other and so producing rupture of the fibre with small stresses. In fact there are several, whose relative importance varies from fibre to fibre.

The first inter-chain force is entanglement. In an unoriented polymer, where the chains are all coiled, it is easy to envisage the tangled mass of chains intertwining among each other. In the fibre there is more order, but still individual chains can have long oriented sections succeeded by twisted, intertwined sections and if they are long enough (hence the minimum molecular weight—see p. 10) each chain will contain sufficient intertwined sections to make it difficult to separate from its neighbours. This type of force is more effective if the polymer contains a high proportion of rigid units, such as ring structures, than if it consists only of a flexible chain.

The second type of inter-chain force is that exerted between individual atoms or small groups of atoms in neighbouring chains. The strongest force of this type, a covalent bond, is rarely found between chains in fibres for reasons to be explained later. Ionic bonds, less powerful polar bonds such as hydrogen bonds, and other weak inter-atomic forces can all contribute. Individually each bond may be quite weak, yet if many weak bonds connect a single long chain to its neighbours then the total force holding that chain in place will be high.

The third type of inter-chain force is crystallinity, which represents an ordering of the weak forces. Most simple compounds can form crystals in which many individual molecules are arranged in a regular fashion. So too many polymers can crystallize, but with the important difference that instead of entire molecules only parts of molecules are incorporated into crystalline regions. Whereas the simple compound can be regarded as completely crystalline, the polymer is very much less than completely crystalline. It is usual to regard a crystalline polymer as consisting of regions having different degrees of order ranging from perfectly crystalline through partly crystalline regions to completely non-crystalline. A single molecule is much longer than a single crystalline region, so different segments of its chain may lie in different crystalline regions. (Fig. 2.1.) Whenever a molecular chain passes through a highly crystalline region it is bound by the relatively strong forces of crystallinity to each of the neighbouring chains. This third type of inter-chain force is a much longer range force than that arising from interatomic attractions.

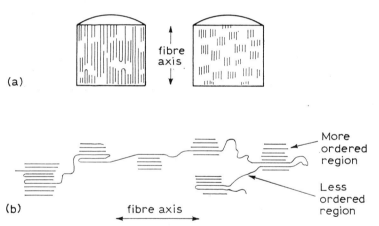

FIG. 2.1   **a** Different views of fibre structure: (l.), ordered structures containing disordered regions; (r.), disordered structure containing ordered regions. **b** How a single molecule can form part of several more and less ordered regions.

Some fibres, such as the acrylics, do not crystallize. Crystallinity is not essential to useful fibre formation.

Methods used to investigate the shape of polymer chains in fibres and the ways in which they are held together include X-ray diffractometry, which provides information about crystallinity and related kinds of order; infrared and nuclear magnetic resonance spectrometry, which show how individual chemical bonds are oriented within the molecule; and electron microscopy, which is able to reveal fibrillar structures within the filaments.

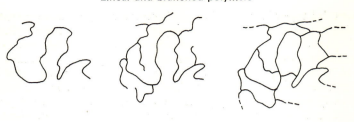

Linear   Branched   Cross-linked

FIG. 2.2   Linearity and branching in macromolecules.

2.2   Linear and branched polymers

Polymer chains may be more or less branched (Fig. 2.2). Linear poly-mers are by far the most important for fibres; cotton, silk, and the major man-made fibres are essentially linear in structure. Linear polymers can be melted or dissolved and can be oriented so that their chains lie more nearly parallel. The more branching is present, the more difficult these processes become. Completely cross-linked polymers have virtually infinite molecular weight and will not process into fibres. Yet cross-linked fibres exist. Wool is naturally cross-linked, and some specialized synthetic fibres are cross-linked after the fibrous shape and orientation have been produced.

2.3   Chemical structure and fibre properties

We are all aware of differences between fibres, and have our own beliefs, and sometimes prejudices, about which fibres are best for various purposes. Wool has a reputation for warmth, cotton for summer coolness; nylon makes good stockings, polyester keeps the crease in trousers. Why? The chemical structure of the fibre nearly always provides the answer. A review of some of the more important fibre properties will show how.

Moisture absorption
Materials that readily absorb water are said to be hydrophilic; those that do not are hydrophobic. From Fig. 2.4 it can be seen that wool is much the most hydrophilic of the common fibres, and that the natural and regenerated fibres are all more hydrophilic than the fully synthetic fibres.
The chemical structures reponsible for high moisture absorption are all relatively polar groups that can associate with water molecules by forming hydrogen bonds, such as those illustrated in Fig. 2.3.
The two most important are the amide group and the hydroxyl group, which account for most of the moisture capacity of wool and cotton respec-tively. Since crystallites do not absorb water, high crystallinity reduces

moisture absorption; the differences between wool and silk, and between viscose rayon and cotton, are partly due to the lower crystallinity of the former in each pair. In more hydrophobic fibres hydrogen-bonding polar groups are infrequent or totally absent.

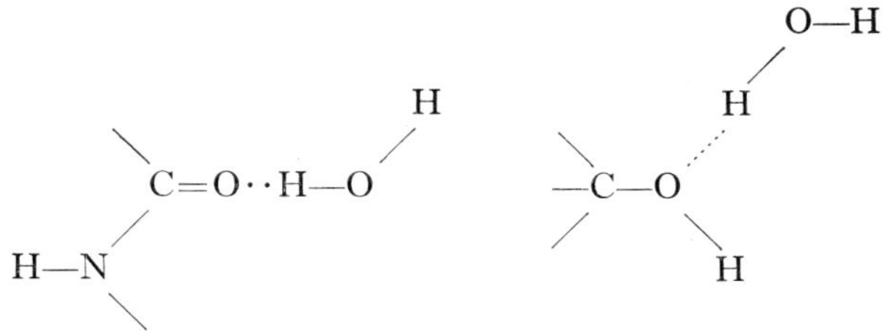

FIG. 2.3   Water molecules associated through hydrogen bonds with amide (l.) and hydroxyl (r.) groups.

The early progress of synthetic fibres was largely due to easy-wash, easy-dry properties that natural fibres lacked. Cotton creased, wool shrank. Here their high moisture absorption proved disadvantageous, for hot water causes a loosening of inter-chain forces that leads to loss of fabric smoothness or shape. Cotton therefore had to be ironed, wool carefully washed. Non-absorbent synthetic fibres are easy to wash, quick to dry, and need no ironing. Nowadays cotton and wool can be, and often are, chemically treated to improve their dimensional stability.

On the other hand, natural fibres have been commonly supposed to be more comfortable than synthetics, because of their higher moisture absorption. There may be some truth in this, particularly under extreme conditions, but fabric structure is far more important. Knitted fabrics differ from woven, bulked yarns from unbulked, staple fibre from continuous filament in handle and in transmission rates for heat and moisture. These differences are used to design comfortable fabrics even from the most hydrophobic fibres.

Softening and setting

There is a well known and spectacular experiment in which a piece of flexible rubber is dipped for a short time in liquid nitrogen, removed, and tapped sharply with a hammer, whereupon it disintegrates into many irregular pieces. Rubber behaves in this way because when cooled it passes through a temperature known as the glass–rubber transition, or T_g for short, below which it is rigid and above which it is flexible. This transition is characteristic of non-crystalline materials: if they are cross-linked (like rubber), they change from a glass to a rubber on passing upward through this temperature; if they are not cross-linked (like polystyrene), they soften and flow above T_g, which then represents a softening point—in the case of polystyrene at about 100 °C. Non-crystalline fibres like secondary cellu-

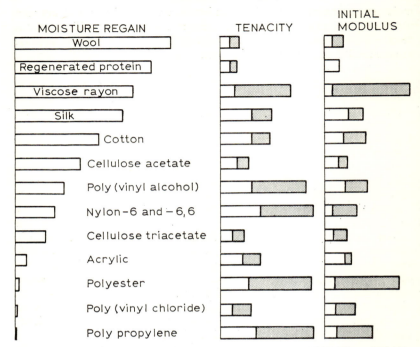

FIG. 2.4   Typical fibre properties: moisture regain at 65% R.H. and 21 °C;
tenacity (strength); and initial modulus (stiffness). Shaded areas
represent range of values.

lose acetate soften and shrink above T_g, which marks the upper limit of
their usefulness.

Fibres that crystallize become only partly crystalline, and their non-
crystalline regions similarly exhibit glass–rubber transitions. The crystallites
act like the cross-links of rubber in preventing the polymer from flow-
ing at temperatures between T_g and the melting point of the crystallites
(T_m). Such fibres do not soften and flow until T_m is exceeded. Nylon-6,6
and most polyester fibres are crystalline and melt at about 260 °C; polypropylene melts at 167 °C. At higher temperatures they become viscous
fluids. Cotton, viscose, and silk are also crystalline, but decompose before
reaching their melting points.

Even in crystalline fibres T_g remains important. Below T_g fibres will
not orient by 'drawing' (Chapter 3), since the polymer chains are unable to
slide over one another into a more extended configuration; nor will they
crystallize because the necessary re-alignment of polymer chains cannot
occur. If T_g is below room temperature it is relatively unimportant: in
polypropylene, for example, and in damp nylon-6,6. But if T_g is above

room temperature it becomes necessary to heat the fibre to orient and crystallize it. Polyester fibres need such treatment. They are oriented at about 90 °C, then heated to about 140 °C while still under tension to crystallize them. Without crystallization they would shrink disastrously when heated in use.

A valuable property of poly(ethylene terephthalate) and cellulose triacetate fibres is their ability to retain pleats, and this property is connected with the fact that the oriented, crystalline fibres have a glass transition temperature well above the temperature of boiling water. If a pleat is formed by pressing the fabric at, say, 150 °C, the polymer chains move about to adopt the positions required to hold the pleat in position and 'freeze' into these positions on cooling below the transition temperature. When the fabric is washed, the temperature does not rise high enough to unfreeze the chains so the pleat remains in place.

Anything that makes it more difficult for polymer chains to move relative to one another raises T_g: many rigid units in the chain; frequent and powerful polar bonds between chains; heavy cross-linking; high crystallinity; high orientation. Plasticizers make it easier for chains to move, so they lower T_g. It is the plasticizing action of water on nylon-6,6 that pushes its T_g below room temperature except under very dry conditions, and so helps creases to drop out of nylon shirts in wear.

Rigid units and polar bonds also favour high melting point, but structural symmetry is perhaps more important still. Take poly(ethylene terephthalate) and poly(ethylene orthophthalate), whose repeat units are I and II respectively.

        (I)   (II)

(I) is symmetrical; the polymer melts at about 260 °C.
(II) is unsymmetrical; the polymer is not crystalline at all.
Another kind of symmetry is also relevant. Pure poly(ethylene terephthalate) consists only of repeat units like (I). If some other repeat unit—say (II)—is also present it cannot be incorporated into the crystallites formed by (I), so the crystallinity and the melting point are reduced.

Mechanical properties

Many of the most important properties of fibres are mechanical: their breaking strength; their extension at break; their stiffness, or resistance to deformation; their recovery from imposed strains. Chemical structure and physical processing both play a part in determining these properties. Strength, for instance, is first and foremost a function of molecular weight. If a strong fibre is degraded by reducing the molecular weight of the polymer, it becomes progressively weaker and eventually falls apart. Sunshine has such a rotting effect on some fibres by providing energy for oxidation of the polymer, and acid or alkali will rot others by hydrolysing the polymer. Below a certain molecular weight a polymer is unable to form fibres; just above that molecular weight it can form weak fibres. As the molecular weight rises, so the strength of the fibres increases. For many synthetic polyamides and polyesters fibre formation becomes possible at a molecular weight of about 5000, and at values above about 10 000 commercial fibres can be made.

High molecular weight alone will not ensure high breaking load. The degree of orientation of the fibre is also crucial. Higher orientation means greater breaking load. It also means lower extension to break. It is therefore possible to alter the balance between these two properties by altering the orientation.

Microscopic imperfections in the fibre, such as voids or solid inclusions, reduce its strength. They reduce the load-bearing area; they concentrate stresses so that cracks are more readily initiated. But they are not to blame for the fact that the actual strength of a fibre falls far below the theoretical strength based on breakage by covalent bond rupture. The imperfections that account for most of the discrepancy are on a molecular rather than a microscopic scale.

Ordinary textile uses do not require the highest possible fibre breaking strength, but certain industrial outlets do. Tyre cords, conveyer belts, ropes, and sewing threads are examples. Fibres for these uses tend to be of higher molecular weight, higher orientation, and greater structural perfection than ordinary textile fibres.

The stiffness, or resistance to deformation, varies very much according to the direction in which the stress is applied. The most important direc-

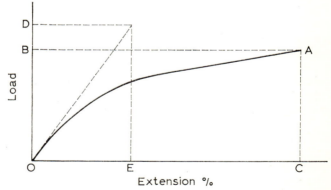

FIG. 2.5   Typical load-extension curve (OA), showing breaking load (B), extension at break (C), and initial modulus (D × 100/E, expressed per unit cross-sectional mass).

tion is along the fibre axis. The measure of stiffness most often quoted is the *initial modulus* for extension (Fig. 2.5). High stiffness is valuable in tyre cords and conveyer belts, and particularly important in composite materials where the fibres act as reinforcement for another, non-fibrous polymer. Use of glass fibre to reinforce polyester resins for making car bodies, boat hulls, and other rigid structures is a typical case. The initial modulus of glass fibre is very high. That of carbon fibre is even higher. These two fibres have rigid cross-linked chemical structures so they have very limited capacity for extension by chain slippage; the immediate resistance to deformation is very high, but the total extension possible is so low that they are brittle when bent. Rather less stiff, but also less brittle, are fibres made from polymers such as poly(ethylene terephthalate) that contain a high proportion of rigid groups, in this case aromatic rings. At the other end of the stiffness scale come the elastic fibres, where low resistance to deformation is achieved by including long flexible chain units in the polymer.

3 Principles of fibre-making processes

3.1 Spinning processes

Bulk polymers have no particular shape and their molecular chains are oriented in no particular direction. Polymers in fibrous form have a well-defined shape and their molecular chains are highly oriented. The transformation from one condition to the other is brought about by the process known as *spinning*, usually supplemented by a further process known as *drawing*. In spinning, the polymer is pushed through fine holes then pulled away rapidly as it emerges so that a fine filament is produced from each hole. Bulk polymers must be converted into viscous fluids before they can be pushed through the holes, so they are either dissolved or melted. Of the three main ways in which fibres are spun from non-fibrous polymer, two, *wet-spinning* and *dry-spinning*, employ solutions of polymers; the third, *melt-spinning*, employs molten polymer. Although melt-spinning is the most recent of the three, it is simplest in principle and will be considered first.

Melt-spinning (Fig. 3.1)

In melt-spinning the polymer is melted and extruded through the fine holes of a spinneret—a metal plate with numerous small holes of around 0.01 inch diameter drilled through it—into a cool atmosphere. The threadline loses heat by radiation and by transfer to a controlled air draught and quickly solidifies. Several yards below the spinneret is a cylindrical roll rotating at a high speed. The bundle of filaments is given a high velocity—several thousand feet per minute—by passing round the roll before being forwarded to a collecting device. This high velocity attenuates the filaments while they are still in the molten state just after emerging from the spinneret holes. The diameter of the fibres collected at this stage depends on the rate at which molten polymer is supplied to the spinneret, which is closely controlled by a gear pump, and on the surface speed of the roll, but not on the diameter of the spinneret holes.

The fibre is not yet suitable for use, because the polymer chains are not highly oriented. To orient them the fibre is subjected to a process called *drawing* in which it is attenuated still further, but this time in the plastic rather than the molten state. Figure 3.2 illustrates one of the many arrangements for drawing fibres. The surface speed of the draw roll is several times that of the feed roll; the ratio of the speeds is known as the *draw ratio*.

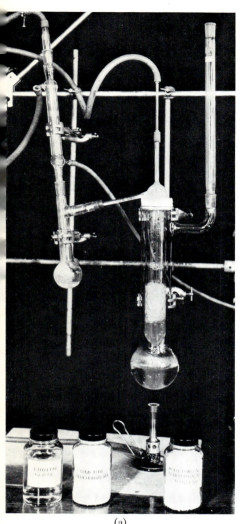

(a)
Laboratory demonstration of melt
polymerization (ICI Fibres Ltd)

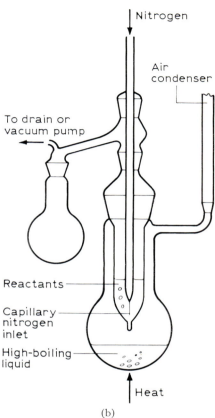

(b)
Glass apparatus for demonstrating melt
polymerization

PLATE I

PLATE 2 Laboratory demonstration of wet-spinning (Courtaulds Ltd)

PLATE 3 Melt-spinning polyester staple (ICI Fibres Ltd)

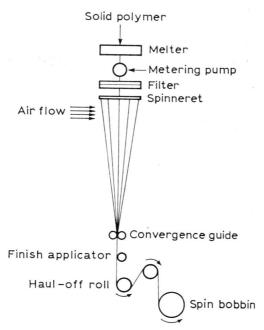

FIG. 3.1   A melt-spinning process.

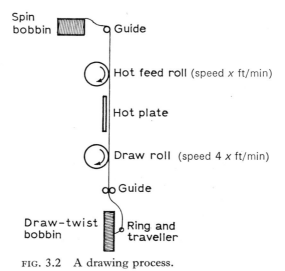

FIG. 3.2   A drawing process.

B

To be suitable for melt-spinning, a polymer must be stable at the melting and extrusion temperatures, which are generally at least 20 °C above its melting point. Four commercial melt-spun fibres are particularly important—nylon-6,6, nylon-6, poly(ethylene terephthalate), and polypropylene.

EXPERIMENT 1. Nylon fibres from a comb.

Touch the surface of a *nylon* comb with a hot (but not red-hot) glass rod and withdraw the rod slowly. With a little practice long filaments of nylon can be produced.

Dry-spinning

Although dry-spinning is the least used of the major spinning techniques, historically it was the first for it was used by Chardonnet in commercial production of his original artificial silk. It depends upon extrusion of a polymer solution through a spinneret into a heated gas or vapour so that the solvent is rapidly evaporated, leaving filaments. There are some obvious limitations to this technique. It depends on the existence of a suitable cheap, volatile solvent for the polymer. If there is a chemical regeneration reaction required to produce the fibre from the solution (as in viscose rayon

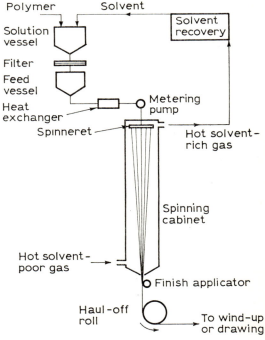

FIG. 3.3   A dry-spinning process.

production) wet-spinning is nearly always preferable because regeneration requires a wet process. On the other hand, if solvent recovery is essential—as it is for all solvents but water—the volume handled and the expense of recovery will be lower for a dry-spinning process than for wet-spinning.

Figure 3.3 illustrates a typical dry-spinning lay-out. Hot air, nitrogen or superheated solvent vapour is passed through the cabinet into which the solution is being extruded and removes solvent from the threadline so that the fibres emerge at the foot of the cabinet nearly solvent-free. Acetone and methylene chloride (dichloromethane), the first important dry-spinning solvents, are used in producing fibres from secondary cellulose acetate and cellulose triacetate. Both these solvents are very volatile. Household polystyrene adhesive is a solution in a similarly volatile solvent, and threads of polystyrene can easily be drawn from its surface at room temperature in a simulation of dry-spinning. Acrylic and spandex fibres, on the other hand, are spun from much less volatile solvents, such as dimethyl formamide and dimethyl acetamide, which boil in the 150–200 °C range and require much higher temperatures to evaporate them.

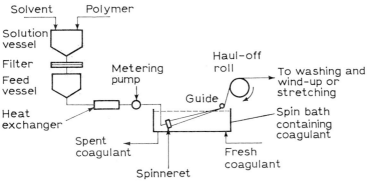

FIG. 3.4   A wet-spinning process.

Wet-spinning

Any soluble high polymer can in principle be wet-spun by extruding it through a spinneret into a non-solvent or precipitant and pulling the freshly precipitated polymer away from the spinneret holes at a controlled rate. The rate at which filaments can be drawn away from the spinneret holes without breaking is much lower in wet-spinning than in melt-spinning or dry-spinning. So for equivalent filament diameters the rate of supply of polymer to each hole is much lower. This lower throughput per hole would be a serious deficiency, but it is more than compensated by another factor. In melt-spinning the distance between holes cannot fall below a figure controlled by the rate at which heat can be removed from the filaments; in dry-spinning the controlling factor is the rate at which heat can be provided to the filaments to evaporate the solvent. In wet-spinning, however, there is very much less limitation due to heat transfer, partly because there is little change in temperature during extrusion, partly because the liquid coagulant bath is a much better heat transfer agent than the gas or vapour used in the other processes. So the holes in a spinneret used for wet-spinning can be closer together, and the number in any one spinneret can be much greater—so much so that the throughput per spinneret can be higher than in melt- or dry-spinning.

Wet-spinning processes fall into two broad classes, (a) those that employ a solvent/non-solvent system, and (b) those that employ a chemical reaction to dissolve and precipitate the polymer. Fully synthetic fibres such as the acrylics and modacrylics are usually spun by the former class of process, whereas regenerated natural fibres such as viscose and cuprammonium rayons require the second class. Examples of both are discussed in more detail at a later stage.

EXPERIMENT 2. Wet-spinning in the laboratory.

(a) From a hypodermic syringe.

Prepare a clear viscous solution of the polymer. Half-fill a hypodermic syringe with the solution and ensure that no bubbles are present. Immerse the tip of the needle in a narrow coagulant bath about 2 feet long. Press firmly so that polymer solution emerges from the needle. As it coagulates grasp it in a pair of tweezers and pull gently away while continuing to inject solution into the bath. With a little practice, long filaments will readily be produced.

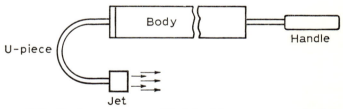

FIG. 3.5   Garden syringe modified for laboratory wet-spinning.

(b) From a garden syringe.

Modify a garden syringe as shown to connect the multi-hole jet to the body by a U-piece. Fill the syringe with clear polymer solution, avoiding air bubble formation. With the syringe upright in a bucket of coagulant push the plunger. Grasp the filaments as they form at the jet surface and pull them vertically out of the coagulant while continuing to press the plunger. Two persons are needed; the filaments can be collected on a rotating drum operated by a third person.

(c) Suitable materials.

Sodium alginate is a suitable polymer—see p. 30. Other suitable polymer–solvent–coagulant systems are: polyacrylonitrile–dimethyl formamide–aqueous dimethyl formamide; poly(vinyl alcohol)–water–aqueous sodium sulphate. In these latter cases the polymer concentration in spinning solvent and the coagulant composition should be varied in order to identify good conditions and learn what conditions are unsuitable. This can conveniently be done on the hypodermic scale before embarking on the garden syringe experiment. Observe what goes wrong at very high and very low polymer and coagulant concentrations.

TABLE 1   How some man-made fibres are spun

Melt-spun	Dry-spun	Wet-spun
Nylon-6,6	Secondary cellulose acetate	Viscose rayon
Nylon-6	Cellulose triacetate	Cuprammonium rayon
Polyester	Acrylic (e.g. Orlon)	Acrylic (e.g. Courtelle, Acrilan)
Polypropylene	Spandex (e.g. Lycra)	Spandex (e.g. Vyrene)
	Poly(vinyl chloride)	Poly(vinyl chloride)
		Protein
		Poly(vinyl alcohol)
		Aromatic polyamides

3.2   Filament yarn and staple

The fibre-forming processes just described give virtually endless filaments. Of the natural fibres only silk is of this type. All the others are available only in short lengths, because that is how they grow on a plant or animal. The short lengths are known as staple fibre, or staple, and the continuous filaments as filament yarn. Although this distinction has no chemical significance, it is very important in textile processing, where staple fibre undergoes a number of extra stages to convert it into a yarn, and in textile products, where the two kinds of yarn provide different fabric aesthetics. It took over twenty years for the producers of man-made fibres to appreciate the value of converting their continuous filaments into staple by chopping them up into short lengths similar to the natural length of cotton or wool. The textile processing then became more complex; in return it became possible not only to use man-made fibres to make fabrics with staple aesthetic qualities, but also to mix man-made fibres with natural fibres and so to obtain new and useful blend fabrics.

It might be expected that the extra textile processes would make the staple fabric more expensive. In practice this is more than compensated by an increase in productivity in the fibre manufacturer's process. In spinning filament yarn the number of filaments extruded from each spinneret is limited because each bundle of filaments must be collected separately. When spinning staple the number of filaments extruded from each spinneret is limited only by the practical weight of the spinneret assembly and by the spacing between holes needed to avoid coalescence of filaments. Bundles of filaments from many spinnerets may be combined for subsequent processes to form a single bundle containing hundreds of thousands of filaments and known as a *tow*. Because one unit of machinery can handle so much more fibre at a time, staple is cheaper to produce than filament yarn.

These factors explain why man-made staple production rapidly grew so that it soon exceeded filament yarn production, as it still does. The cost difference remains, but the aesthetic difference no longer favours staple. Many processes have been developed for *bulking* filament yarns so that each multifilament bundle occupies a much greater volume that it would if the individual filaments remained straight and virtually parallel (Fig. 3.6).

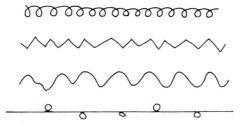

FIG. 3.6   Four of the shapes that individual filaments in a bulked yarn may adopt.

The products are well known under such trade names as Ban-Lon, Crimplene and Helanca, which describe just three of the wide variety of bulked filament yarns now available.

4 Fibres from natural polymers

4.1 Natural cellulosic fibres

By far the most important single textile fibre is cotton, the seed hair of the cotton plant, which is cultivated in most sub-tropical areas of the world for the sake of the fibre. Chemically it consists predominantly of the polymer *cellulose* (I), the main building block of the vegetable world, and the basis of several important natural fibres beside cotton. The best known are flax (the source of linen), jute (widely used in sacks), Manila hemp and sisal hemp (used in ropes). It is easy to forget, however, that fibres have other uses besides fabrics and ropes, notably paper. At least five times as much cellulosic fibre is used in paper as in fabrics; in this use cellulose remains unchallenged, apart from a few very specialized products, to this day. The fibres used for making paper are so short and irregular in shape that they are useless for making fabrics by the old, well-established methods. They are cheaper than cotton because they are produced from wood by pulping it with chemicals that dissolve away the lignin and liberate the short cellulosic fibres.

Cellulose is a member of the important, complex class of chemical materials known as carbohydrates, because their molecular formulae can be written as if they consisted of carbon and water only. The sugars glucose (II) and sucrose are relatively simple examples, with molecular formulae $C_6H_{12}O_6$ and $C_{12}H_{22}O_{11}$ respectively, of the two classes known as monosaccharides and disaccharides. By the middle of the nineteenth century it was recognized that cellulose was a polysaccharide $(C_6H_{10}O_5)_n$, because boiling it with dilute acid gave one of the two forms of glucose. In 1922 Irvine and Hirst proved that it gave only glucose. They went on to show that each of the glucose units in cellulose contains three hydroxyl groups, and worked out which of the five hydroxyl groups present in uncombined glucose these three were. After several years' vigorous argument, carbohydrate chemists finally agreed that the glucose units were present as six-membered rings, and that the cellulose repeat unit contained two glucose units arranged as in (I). More correctly they are glucose minus water, or anhydroglucose units.

The concept of linear macromolecules had only just been established by Staudinger and others, and it took some years for the structure (I) to be fully accepted. The isolation in 1932 of a little tetramethylglucose (VII) from the hydrolysis product of fully methylated cellulose, and the recog-

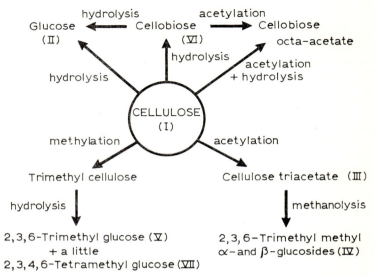

FIG. 4.1   Steps in the determination of the structure of cellulose.

nition that this arose from an end-group (VIII) of the polymer molecule, helped to carry conviction. Formula (I) is universally accepted today.

The next question is the molecular weight of the cellulose polymer. All chemical and solution treatments of cellulose degrade it to some extent.

I cellulose repeat unit

II a

α   II b   β

II glucose: (a) open chain structure
              (b) cyclic (pyranose) structure; α-form
                  and β-form are freely interchangeable

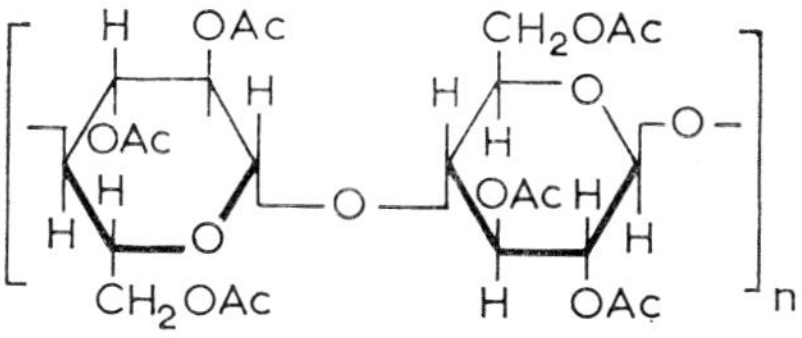

III cellulose triacetate (Ac = CH₃CO−)

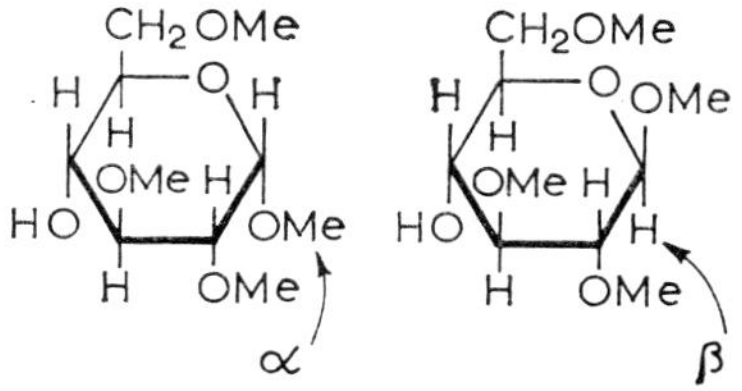

IV 2,3,6−trimethyl methyl α−and β−glucosides; these α−and β−forms are methyl ethers therefore not freely interchangeable

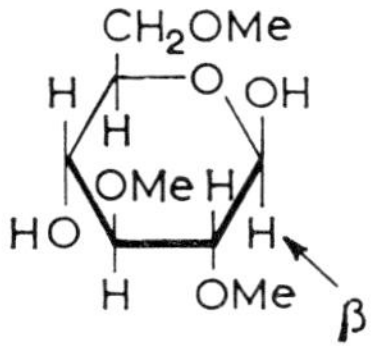

V 2,3,6−trimethyl glucose (α−form ⇌ β−form)

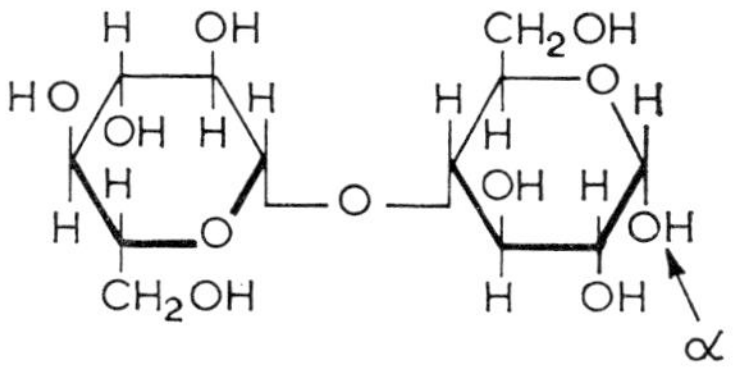

VI cellobiose (α−form ⇌ β−form)

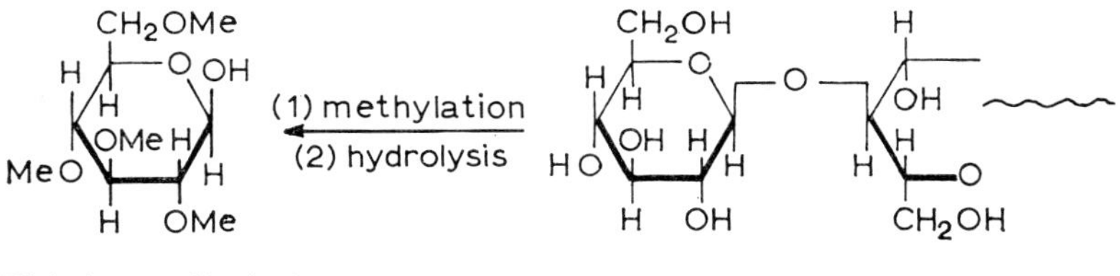

(1) methylation
(2) hydrolysis

VII tetramethyl glucose (α−form ⇌ β−form)

VIII cellulose end-group

Moreover, celluloses from different sources may have different molecular weights, so it is difficult to answer with certainty. The consensus of opinion is that cotton cellulose has a molecular weight above 500 000. It contains at least 3000 anhydroglucose units on average in the polymer chain, and possibly many more.

Cellulose can be obtained in more than one crystalline form, the most important being known as Cellulose I and Cellulose II. Cellulose I is the form that occurs in the natural fibres; cellulose that has been regenerated from a soluble derivative is in the Cellulose II form. Vigorous swelling of cotton, as for instance in the process called mercerization in which cotton is treated with 20–25% caustic soda for a short time to increase the dyeability and moisture absorption and give it a glossy appearance, also converts it from Cellulose I into Cellulose II.

Crease resist and permanent press treatments

Cotton and the regenerated cellulose fibres, such as viscose rayon, have little resistance to the development of creases. The development of methods of improving their crease resistance, and latterly also their retention of imposed creases and pleats, has been an important factor in maintaining the importance of the cellulosic fibres in the face of competition from the newer synthetic fibres.

The first important development was in 1926, when Foulds, Marsh and Wood found that if urea-formaldehyde condensates of low molecular weight were used to impregnate cotton and rayon fabrics, then the fabrics were 'cured' by heating with an acid catalyst at a high temperature, crease-resisting properties were produced. They showed that the resin must penetrate the fabric, not just cover the outside. At one time it was believed that deposition and polymerization of resin in the interstices caused the crease-resistant effect. It is now generally believed that the principal mechanism is cross-linking of the cellulose by covalent bonds:

$$\text{OH HO CH}_2\text{ NH CO NH CH}_2\text{OH HO} \xrightarrow[\text{150-200°C}]{\text{catalyst}} \text{O CH}_2\text{NH CO NH CH}_2\text{O}$$

cellulose	N, N'-bishydroxymethyl-urea	cellulose	cross-linked cellulose

The effect of the cross-links is to produce a three-dimensional network such that the fibre when stressed 'remembers' its previous configuration and tends to return to it. Very many other cross-linking agents have since been developed, but the most important step forward has been the recent introduction of delayed cure processes. The fabric is impregnated with a pre-polymer, then the garment is made up before curing. The curing process is carried out with the garment in the shape finally required—for example, a pair of trousers is cured by heating in a press with the crease in the required place—so the garment when deformed in later wear or washing tends to return to the shape given to it during the curing. This type of process has become known as 'permanent press', and its success depends very much on the choice of cross-linking agent and of curing catalyst. A combination which does not give premature cross-linking is essential. The bishydroxy- and bishydroxymethyl-imidazolidin-2-ones are particularly useful, notably 1,3-bishydroxymethyl-4,5-dihydroxyimidazolidin-2-one (dimethylol dihydroxy ethylene urea; DMDHEU):

DMDHEU

EXPERIMENT 3. 'Permanent press' cotton.

Take two pieces of a cotton sateen fabric each about 1 foot square. Leave one untreated. Dip the other in an aqueous solution containing 10% DMDHEU   (1,3-bishydroxymethyl-4,5-dihydroxyimidazolidin-2-one) and 1% zinc nitrate hexahydrate, and mangle it lightly so that the wet fabric weighs 180–200% of its original dry weight. Dry the fabric by hanging it up at a temperature not above 60 °C. Pleat it by pressing or ironing at a temperature in the range 100–160 °C, then heat the pleated fabric at 160 °C for 10 minutes.

Iron the untreated sample to introduce pleats in the same way, then wash both samples in hot water plus soap or detergent, and spin dry or hang up to dry. Note that the treated sample retains its pleats and recovers from creases introduced during washing much better than the untreated sample.

For simplicity this experiment omits the addition of softeners, which are flexible polymers normally applied at the same time to prevent the fabric from becoming too stiff and harsh.

4.2 Man-made fibres derived from cellulose

The proportion of cellulosic material that occurs naturally in a form suitable for textile use is very small, so cotton is more expensive than the other, less useful, forms of cellulose. This prompted development of methods of upgrading cheaper celluloses into useful textile fibres. There are two major sources of cheaper cellulose. One is wood pulp, which is also used on an enormous scale for paper manufacture. The other, linters, is cotton too short for textile use.

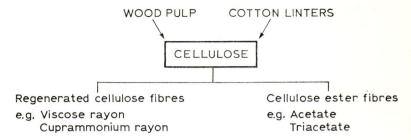

WOOD PULP   COTTON LINTERS

CELLULOSE

Regenerated cellulose fibres   Cellulose ester fibres
e.g. Viscose rayon   e.g. Acetate
      Cuprammonium rayon   Triacetate

Viscose Rayon

The first major man-made fibre, viscose rayon, was developed by Cross and Bevan, who discovered the process in 1891. It is wet-spun from an aqueous solution into an aqueous coagulating bath. The process is chemically more complex than for any other fibre. Solution depends upon reaction of the hydroxyl groups of cellulose with carbon disulphide in the presence of alkali. This reaction is a general reaction of alcohols. The product is structurally a half ester, half salt of the unstable dibasic acid dithiocarbonic acid and is known as a xanthate:

$$CH_3OH + NaOH + CS_2 \rightarrow CH_3O\overset{\overset{S}{\|}}{C}S^- Na^+ + H_2O$$

methanol   sodium   carbon   sodium methyl   water
             hydroxide   disulphide   xanthate

$$)\!\!-\!OH + NaOH + CS_2 \rightarrow)\!\!-\!O\overset{\overset{S}{\|}}{C}S^- Na^+ + H_2O$$

cellulose   sodium cellulose
                                       xanthate

There are several stages between the wood pulp and the spinning solution, or 'dope'. Wood pulp is first reacted with an excess of strong (about 18%) sodium hydroxide to give alkali cellulose, which is still insoluble in water.

$$)\!\!-\!OH + NaOH \rightarrow)\!\!-\!ONa + H_2O$$

cellulose   sodium cellulose
                         xanthate

Excess alkali and some alkali-soluble impurities are removed by pressing them out of the solid. The wet cake is shredded to fine particles known as crumb, then allowed to age in the presence of air for a period of 20–40 hours together with a small amount of a salt of a heavy metal, such as manganese, that acts as an oxidation catalyst. Oxidation during ageing breaks some polymer chains, so the molecular weight of the cellulose falls from about 130 000 to about 60 000.

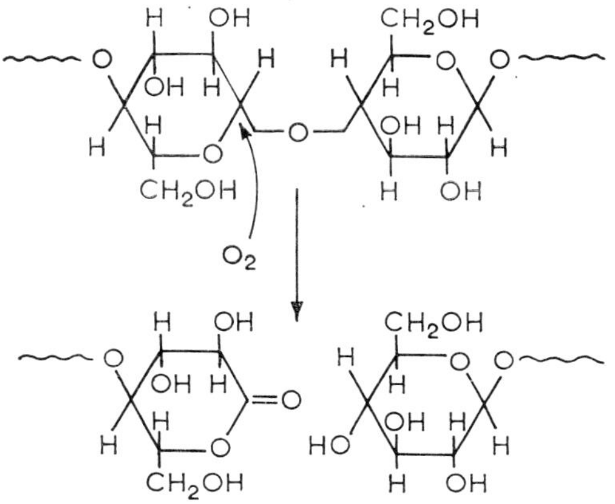

A cellulose chain being broken by oxidation.

The product is now mixed thoroughly with carbon disulphide, CS_2—a low-boiling, toxic, flammable compound, so an air-tight vessel must be used to prevent carbon disulphide from escaping and air from getting in. The reaction produces sodium cellulose xanthate, a polymeric salt derived from cellulose. About one hydroxyl group in every six has now been converted into a xanthate group.

(A) —(AHG)—(AHG)—(AHG)—(AHG)—(AHG)—(AHG)—(AHG)—(AHG)—
      | | | |
     X   X   X   X

(B) —(AHG)—(AHG)—(AHG)—(AHG)—(AHG)—(AHG)—(AHG)—(AHG)—
     | | | |
     X   X   X   X

FIG. 4.2   Sodium cellulose xanthate chains with block (A) and random (B) distributions of xanthate groups (X) on anhydroglucose units (AHG).

Sodium cellulose xanthate is next dissolved in dilute caustic soda in mixing vessels. Although apparently clear the solution is not yet ready for spinning. It must first be *ripened* by storage for a period of several days. The viscosity of the solution first drops due to re-distribution of the xanthate groups, because the blocks of unxanthated units (as in A), which associate with blocks in other chains to hold chains together, are broken up (A → B). The viscosity then rises again due to slow hydrolysis of xanthate back to hydroxyl groups, which by increasing the frequency of the unxanthated units again causes more association between segments of chains. Once the desired viscosity has been reached the 'dope' is ready for wet-spinning, and is extruded through a multi-hole spinneret into a coagulating bath (the spin bath).

The coagulating bath has changed considerably with developments in viscose spinning, and may still vary according to the product required. The simplest bath contains sulphuric acid (about 10%) and sodium sulphate (about 20%), which together precipitate the insoluble and unstable acid from the sodium cellulose xanthate. The acid is so unstable that it rapidly eliminates carbon disulphide, and reverts to cellulose:

$$
\begin{array}{ccc}
\underset{\text{salt}}{\overset{\displaystyle S}{\underset{\displaystyle\parallel}{)\!-\!O\!-\!C\!-\!S^-\,Na^+}}} \xrightarrow{\ H^+\ } & \underset{\text{acid}}{\overset{\displaystyle S}{\underset{\displaystyle\parallel}{)\!-\!O\!-\!C\!-\!SH}}} \rightarrow & \underset{\text{cellulose}}{)\!-\!OH} + CS_2
\end{array}
$$

Some complex side-reactions also occur and give by-products that are even more noxious than carbon disulphide, so the atmosphere about a viscose spin bath is very thoroughly vented away from the operators to avoid toxic and fire hazards.

More sulphuric acid and sodium sulphate in the bath means faster coagulation. However, if coagulation is too fast the properties of the yarn will be impaired. If the acid concentration is kept sufficiently low the fibre can be stretched up to 200% after leaving the bath. The highly stretched yarn then has higher strength than normal viscose, which undergoes only about 100% stretch at this stage. In addition, zinc sulphate (1–4%) is nearly always present in the spin bath. Its function is to modify the regeneration process by forming a temporary cross link. The divalent zinc cation Zn^{++} associates with two cellulose xanthate anions:

$$
\overset{\displaystyle S}{\underset{\displaystyle\parallel}{)\!-\!O\!-\!C\!-\!S^-}}\ Zn^{++}\ {}^-S\!-\!\overset{\displaystyle S}{\underset{\displaystyle\parallel}{C}}\!-\!O\!-\!)
$$

The more zinc salt is present, the more the fibre changes from its usual 'skin-core' structure, where the skin is more highly oriented than the core, to an 'all-skin' structure where the whole fibre is of relatively higher orientation. The formation of stronger fibres with a smoother cross-section is further encouraged by including modifiers, such as polyethylene glycol or amines of various kinds, in the viscose dope. These slow down acid penetration and thus control the rate of regeneration. After coagulation and stretching, viscose yarn undergoes a series of washes to purify it.

Cuprammonium rayon

Cellulose can be dissolved in strong aqueous cuprammonium hydroxide. The resulting solution is used to make cuprammonium rayon, a fibre which resembles viscose in having the chemical structure of cellulose. Neither the copper salt nor the ammonium salt of cellulose is soluble—the complex deep blue cuprammonium cation $[Cu(NH_3)_4]^{++}$ is essential. Cellulose is dissolved in a mixture of aqueous ammonia, caustic soda and copper sulphate to give a solution containing about 10% of cellulose, 4% of copper, and 29% of ammonia, which in contrast to viscose needs no ripening. The solution is spun through relatively large spinneret holes into running water, which precipitates fibres of cellulose. The fibres are highly stretched while still plastic, pass through a sulphuric acid bath that removes residual copper and ammonia, and are then washed and dried.

Cuprammonium rayon is much less important now than viscose, and is no longer produced in the U.K.

EXPERIMENT 4.

Dissolve 36 g copper hydroxide and 6 g copper (I) chloride in 3 litres 25% ammonia. Chop 10 g cellulose filter paper finely, wet out in boiling water then collect and remove excess water. Add the filter paper to the cuprammonium solution and stir until dissolved, keeping the solution under nitrogen throughout to prevent oxidation. Filter the viscous solution through glass wool under nitrogen.

Bubble air or oxygen through a portion of the solution for an hour. Compare the viscosity with that of the original solution.

Wet spin the original solution (for equipment see page 16) into water. Run the fibre through or dip it in dilute sulphuric acid, then water, then allow to dry.

Acetate

When cellulose is esterified with an excess of an acetylating agent, all three hydroxyl groups in each anhydroglucose unit react. The product is cellulose triacetate or primary acetate. Partial hydrolysis of triacetate, until the number of acetate ester groups per anhydroglucose unit averages 2.3–2.5, gives secondary acetate. This material was used extensively for coating the wings and fuselage of aircraft in the 1914–18 war. Since 1919 it has been used for making fibres.

Most cellulose triacetate for secondary acetate production is made from cellulose by one of two methods: reaction with a mixture of acetic anhydride, acetic acid, and sulphuric acid, or with acetic anhydride, methylene chloride (dichloromethane), and sulphuric acid. With the former mixture, quite a high sulphuric acid content (8–10% based on the cellulose) is needed to ensure that the acetylation product is soluble, and the reaction must be cooled to keep it below about 40 °C otherwise the polymer will degrade. The viscous reaction product contains triacetate in which a few sulphate ester groups are present. Partial hydrolysis to secondary acetate is carried out by adding the right amount of dilute acetic acid and allowing hydrolysis to occur over several hours. The sulphuric acid is then neutralized with sodium acetate or magnesium acetate. Secondary acetate is precipitated with water, filtered off, and dried.

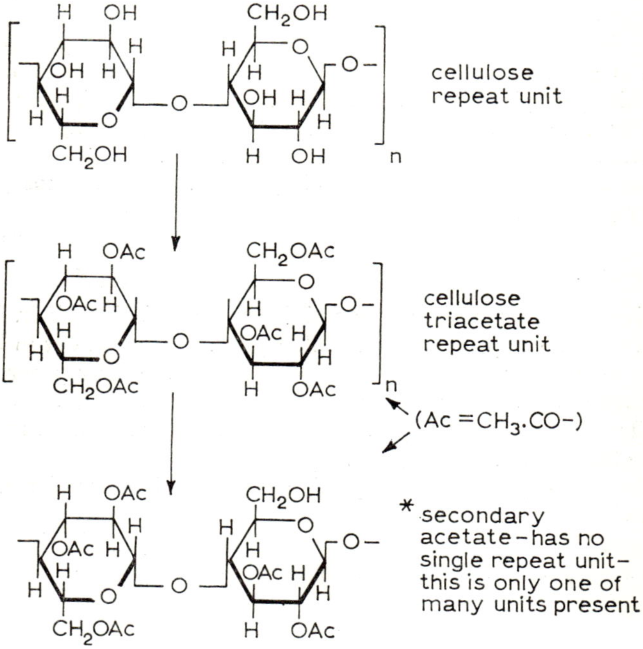

cellulose repeat unit

cellulose triacetate repeat unit

(Ac $=CH_3.CO-$)

* secondary acetate – has no single repeat unit – this is only one of many units present

The methylene chloride process has some advantages. Since methylene chloride is a solvent for cellulose triacetate, less sulphuric acid may be used; since it is low boiling it prevents the reaction mixture from becoming too hot. The methylene chloride must be distilled off after hydrolysis of the triacetate and before precipitation of secondary acetate.

Fibres are produced by a dry-spinning process using acetone as solvent. Secondary acetate, dissolved in acetone containing up to 5% of water to give about 25% concentration of polymer, is extruded into hot air. The filaments are stretched during passage from the spinneret to the wind-up; most acetate fibre is used without any further drawing or stretching stage.

PLATE 4 A filament yarn drawing area (ICI Fibres Ltd)

PLATE 5 The surfaces of *top* wool (× 1300), *centre* viscose (× 4400), and *bottom* polyester (× 4000) fibres. (ICI Fibres Ltd)

Triacetate

We have already seen that complete acetylation of cellulose introduces three acetate ester groups into each anhydroglucose unit. The product, cellulose triacetate, is soluble in partly chlorinated hydrocarbons such as chloroform (trichloromethane) and methylene chloride (dichloromethane). Small amounts of fibre were produced commercially from chloroform solution in the United States during the period 1914–1927. The cost and toxicity proved too high and no suitable dyes were available. Nowadays methylene chloride, CH_2Cl_2, is cheap and readily available. It is also relatively non-toxic, and has been the solvent generally used for spinning triacetate fibres since their re-introduction in 1954.

The cellulose triacetate polymer used for triacetate fibre may be made in the same way as that used as an intermediate for secondary acetate but it must be purified before use. Some sulphate ester groups are present, and need to be removed before spinning to improve the fibre stability. One way is to heat the acetylation product, while it is still dissolved in the reaction mixture, with aqueous magnesium acetate to replace the sulphate groups by acetate or hydroxyl:

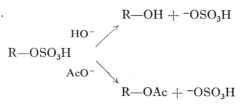

$$R\text{—OH} + {}^-OSO_3H$$
$$HO^-$$
$$R\text{—OSO}_3H$$
$$AcO^-$$
$$R\text{—OAc} + {}^-OSO_3H$$

Triacetate polymer is precipitated by water, dried, then dissolved in methylene chloride to give about a 20% solution. The fibre is dry-spun from this solution into hot air. Because orientation is produced during spinning, the fibre does not need a further drawing step.

EXPERIMENT 5. Acetylation of cellulose.

Chop 60 g cellulose filter paper finely and allow it to stand overnight in 20 cm³ glacial acetic acid. Add 150 cm³ acetic anhydride, 220 cm³ methylene chloride, and 0.34 cm³ 96% sulphuric acid, keeping the temperature below 25 °C. Heat to 40 °C during 1 hour, and keep at 40 °C until the fibre has completely dissolved (check by putting a drop between glass plates and looking through). Add 2 g sodium acetate dissolved in 150 cm³ 50% acetic acid and stir well to destroy the excess of sulphuric acid. Distil off the methylene chloride under reduced pressure, filter the residue, then add it to a large amount of water with vigorous stirring. Cellulose triacetate precipitates; isolate it, wash well with acid-free water, and dry for 24 hr at 60 °C.

Check the solubility in methylene chloride (soluble) and acetone (insoluble).

C

Triacetate differs markedly from secondary acetate in several ways. The most important are its hydrophobic nature and its ability to crystallize. Secondary acetate is structurally a copolymer, containing both triacetate and diacetate groups, so it is not surprising that it is at most very slightly crystalline. Triacetate, on the other hand, contains very few diacetate units, and one result is that it can be persuaded to crystallize by heating it.

Other regenerated cellulose fibres

It is quite possible to convert cellulose ester fibres back to cellulose without destroying the fibrous structure. Chardonnet's first man-made fibre was spun by extruding a solution of cellulose nitrate in a mixture of ethanol and ether into water (wet-spinning) or into air (dry spinning). Because cellulose nitrate is very flammable, the fibres were reduced by ammonium sulphide to cellulose.

Chardonnet's fibre did not long survive the discovery of the viscose process, but an analogous device is now used to produce the extra-strong cellulose fibre known as Fortisan. Secondary acetate fibres are stretched in high-pressure steam, then hydrolysed with dilute alkali to regenerate cellulose. This is, of course, a relatively expensive way of making a cellulose fibre, and is used only for certain special end-uses, such as tyre cords.

Alginate

An interesting, though relatively unimportant, fibre is made from salts of alginic acid, a polymer closely related to carbohydrates which occurs in brown sea-weeds and is extracted from them by treatment with alkali. Alginic acid differs from the polysaccharides such as cellulose in having carboxyl groups, —COOH, instead of hydroxymethyl groups, —CH$_2$OH, attached to the rings.

The sodium salt is soluble in water. Its aqueous solution is spun into a coagulating bath containing slightly acid calcium chloride solution. The calcium salt is insoluble in water, so filaments of calcium alginate form. These fibres are soluble in dilute alkali, which makes them useless as textile fibres but leads to another, very specialized use. They are used as temporary stitches connecting socks during manufacture, and as a temporary framework for lace, and when no longer needed are simply removed by washing in an alkaline soap solution.

EXPERIMENT 6. Alginate fibres.

Mix 5 g sodium alginate with a little cold water to make a stiff dough, then gradually let it down with warm water with stirring to a volume of 500 cm^3. Add about 2 g of a water-soluble dye, preferably blue (e.g. Solway Blue BN) and stir till homogeneous. This is the ***dope***.

Dissolve 500 g calcium chloride in about 5 litres water and add 15 cm^3

concentrated hydrochloric acid. If the solution is hazy filter to clarify. This is the **coagulant**.

Fill the syringe described on p. 16 with the alginate solution, avoiding the formation of air bubbles. Allow the syringe to stand upright for a few minutes to remove air bubbles. With the syringe upright in a beaker of coagulant, push the plunger. As the solution emerges from the jet into the coagulant fibres form. A second person should grasp the emerging fibres (wear a rubber glove) and pull them steadily vertically out of the bath while the first continues to press the plunger. These are alginate fibres.

Alternatively the experiment may be carried out on a smaller scale and the hypodermic spinning method of p. 16 adopted.

Place some of the alginate fibre in dilute sodium hydroxide. Note that it dissolves.

4.3 Protein fibres

Just as carbohydrate is characteristically the basis of the natural vegetable fibres, so proteins are the basis of the animal fibres. Proteins consist of a large number of units linked together by amide (—CO—NH—) groups, so they are in fact a special class of polyamide. Their special nature derives from the fact that the main polymer chain contains only one carbon atom between neighbouring amide groups. However, different carbon atoms along the chain may carry quite different substituents. The proteins may, therefore, be regarded as polyamides derived from a number of different α-amino-acids. The repeat units are known as *peptide* groups; the polymers are sometimes referred to as *polypeptides*.

$$\overset{\displaystyle H}{\underset{\displaystyle R}{H_2N-\overset{|}{\underset{|}{C}}-COOH}}$$
α-amino-acid

$$\overset{\displaystyle H}{\underset{\displaystyle R}{-NH-\overset{|}{\underset{|}{C}}-CO-}}$$
peptide group

In the above formulae the substituent R may be any of a wide variety of groups. The simplest α-amino-acid, glycine, has $R=H$. Other important α-amino-acids have hydrocarbon side-chains; in yet others the side-chains contain polar or reactive groups such as carboxyl (acidic), amino (basic), and hydroxyl- or sulphur-containing groups. Proteins vary in the proportions and sequence of the different peptide units present; no simple repeat unit formula suffices to define their chemical structure. For most proteins, including those of animal fibres, the exact proportions and sequence are still uncertain.

The two important classes of animal fibre are on the one hand hair and wool, and on the other hand silk. Although both are proteins, the chemical differences between them are clear-cut and significant.

Wool

Sheep's wool, the most important of the hair fibres, consists mainly of a round or elliptical central section, or cortex, with an irregular scaly covering known as the cuticle. Although the scaly cuticle influences many of the properties of wool—its handle and its ability to felt, for example— the cortex dominates the bulk properties and its chemical structure is better known. Its chief constituent, the protein α-keratin, may contain as many as nineteen different peptide groups, and their proportions have been measured by hydrolysis of wool and quantitative analysis for each of the α-amino-acids produced. They fall into five groups:

TABLE 2

Side-chain of peptide unit	% of such units in wool
Non-polar hydrocarbon group	35
Free carboxyl group	15
Basic nitrogenous group	10
Sulphur-containing group	10
Other polar groups such as hydroxyl, amide, or non-basic amine	30

Three features of this distribution require special comment. First, there are many acidic and basic groups, which react with basic and acidic dyes respectively to give easy dyeability. Second, no one peptide unit makes up more than 12% of the total. The relatively low crystallinity of wool is due to this diversity of peptide units, for synthetic polypeptides derived from a single amino-acid are very highly crystalline. Third, there are peptide units derived from the amino-acid cystine, and this is perhaps the most significant feature of the whole structure.

Cystine is a sulphur-containing amino-acid with the structure:

$$\begin{array}{cc} NH_2 & NH_2 \\ | & | \\ CH-CH_2-S-S-CH_2-CH \\ | & | \\ COOH & COOH \end{array} \qquad \begin{array}{cc} NH & NH \\ | & | \\ CH-CH_2-S-S-CH_2-CH \\ | & | \\ CO & CO \end{array}$$

cystine   two peptide units derived from one cystine molecule

Since it contains two α-amino-acid groups, there are two derived peptide units. These need not be in the same polypeptide chain. Whenever they are in different chains (and this is believed to be usual) the two chains are linked together through covalent disulphide bonds. A single polypeptide chain has a high molecular weight (probably about 60 000) and about 10% of its peptide units are derived from cystine, consequently each chain is linked to several others. The polymer structure is cross-linked, because no individual chain can be separated from another without breaking a co-valent bond. This cross-linked structure prevents wool from dissolving in polar solvents, as it might well do in view of its low crystallinity and high proportion of polar groups but for the disulphide links. It also plays a major part in ensuring that after wool has been swollen by a polar solvent such as water, or deformed by stretching or crushing, it returns to nearly the same dimensions as before.

The disulphide links can be broken by reduction and re-formed by oxidation. This reaction is commercially and socially important in that best known of all fibres, human hair. Most permanent waving is carried out by reducing the disulphide bonds with a reagent such as the ammonium salt of thioglycollic acid, $HSCH_2COOH$, to mercapto (thiol) groups, then oxidizing back to disulphide with a reagent (the 'neutralizer' of home perms) such as sodium perborate:

$$
\begin{array}{c}
| \\
NH \\
| \\
CH-CH_2-S-S-CH_2-CH \\
| \\
CO \\
|
\end{array}
\qquad
\begin{array}{c}
| \qquad | \\
NH \qquad NH \\
| \text{ reduction } | \\
CH \underset{\text{oxidation}}{\overset{\longrightarrow}{\longleftarrow}} CH-CH_2SH \ HSCH_2-CH \\
| \qquad\qquad | \\
CO \qquad CO \qquad\qquad CO \\
| \qquad\qquad |
\end{array}
\qquad
\begin{array}{c}
| \\
NH \\
| \\
\\
| \\
CO \\
|
\end{array}
$$

disulphide cross-link   two mercapto groups, no cross-link

EXPERIMENT 7. Permanent waving.

Take bundles of straight or slightly curled hair and wool fibres, and a bundle of silk chopped to a similar length. Wind each tightly about a glass rod of 3–5 mm diameter so that each bundle is evenly arrayed one fibre thick in a helix. Fix the ends with Durafix or similar adhesive to prevent slippage.

Immerse the bundles for 20 minutes in a solution of thioglycollic acid (7 g in 150 cm³) adjusted to pH 9.2–9.5 by addition of ammonia. Remove, rinse with water, then immerse in a solution of sodium bromate (15 g in 100 cm³) for 20 minutes. Rinse with water, dry at up to 60 °C, and remove from the wire. Note that the hair and wool are both highly curled and return to the curled form after stretching, whereas the silk is little affected by the treatment. This is because hair and wool both consist largely of keratin, which contains disulphide cross-links, whereas silk fibroin contains no disulphide.

Most permanent waving of hair, and some durable pleat processes for wool, are based upon reduction of disulphide cross-links to mercapto groups then re-oxidation to disulphide while keeping the fibre or fabric in the desired shape. The same reaction can also be used for straightening curly hair.

The polymer chains in wool are not very highly crystalline or very highly oriented along the fibre axis. The main chain of the protein adopts a helical form. In wool this helix is relatively tight and unextended, with much hydrogen-bonding between groups in the same chain. The unextended helix is known in protein chemistry as the α-helix, and the wool protein as α-keratin. When the helix is open and extended, with much hydrogen-bonding between groups in different chains, it is known as a β-helix. Wool keratin can be converted into the β-form by stretching it while wet, but it reverts to the α-form on relaxing. The protein of silk, on the other hand, is in the extended, β-form.

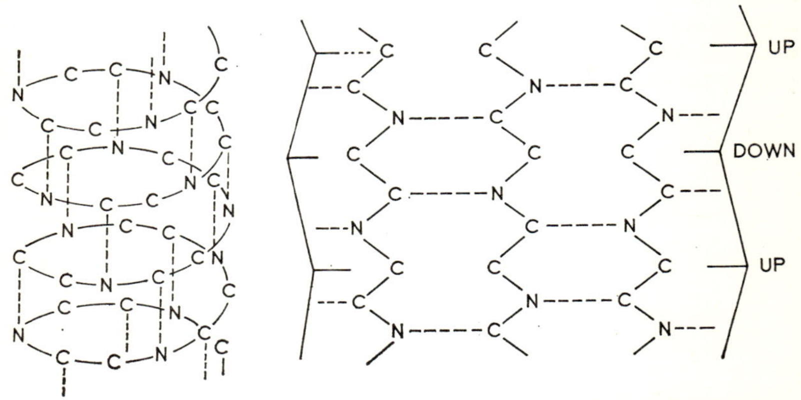

The helical structure of an α-polypeptide. The dashed lines C----N represent *intra*-chain hydrogen) bonds C = O---H—N) between amide groups.

The 'pleated sheet' structure of a β-polypeptide. The dashed lines C----N represent *inter*-chain hydrogen bonds (C = O---H—N) between amide groups.

FIG. 4.3

Silk

The most costly of the natural fibres, for centuries a by-word for luxury, silk makes up for its relative unimportance as a bulk textile fibre by its importance in the development of man-made fibres. Whereas cotton and wool are produced in short lengths, silk is produced in continuous filament form; whereas the polymers of cotton and wool are formed *in situ*, so that the fibre grows by biochemical processes which cannot yet be reproduced synthetically, the silk polymer is non-fibrous when first formed and is converted into a fibre by a spinning process which inspired the early development of man-made fibres.

Many spiders and caterpillars produce silk, the former as a means of catching food or of transport, the latter as a cocoon to protect them when they turn into chrysalids. The silkworm, a moth caterpillar, is the most important. From glands in its head it extrudes two filaments of the protein fibroin, coated and stuck together by another protein, sericin. Industrially, most of the sericin is removed by boiling the fabric in soap solution once its protective action is no longer needed. The final fabric consists almost entirely of fibroin.

Fibroin contains sixteen different peptide units, all of which are also present in wool, but completely lacks the side-chain amide groups and sulphur-containing units present in wool. It therefore has no covalent cross-links. Hydrolysis shows that over 90% of the peptide units are derived from four amino-acids, glycine, alanine, serine and tyrosine, and about 75% from two, glycine and alanine.

$$-NH-CH_2-CO- \qquad -NH-\underset{\underset{CH_3}{|}}{CH}-CO- \qquad -NH-\underset{\underset{CH_2OH}{|}}{CH}-CO- \qquad -NH-\underset{\underset{\underset{\bigcirc-OH}{|}}{CH_2}}{CH}-CO-$$

Peptide units derived from, left to right, glycine, alanine, serine and tyrosine

Partial hydrolysis to di- and tri-peptides and degradation by enzymes have both shown that quite long alternating sequences of glycine and alanine are present. This high degree of chemical order explains the relatively high crystallinity. Since fibroin is linear it can be dissolved, but only in very powerful solvents such as concentrated aqueous solutions of inorganic thiocyanates.

The silk-worm needs no such powerful solvent, because it synthesizes the polymer in the globular, or coiled, form. This form is soluble in water, and the aqueous solution is metastable: applied energy, such as heat, will convert it into the insoluble, extended, fibrous form. The silk-worm applies viscous shear forces during extrusion of the aqueous solution, and thus precipitates oriented fibroin.

There is some uncertainty about the molecular weight of silk fibroin. Values of $\overline{M}_n$ in the range 50 000–300 000 have been reported. These are well above the minimum for fibre formation already quoted, similar to the polypeptide chain length in wool, and below the levels attained in cotton.

The differences between wool and silk are summarized in Table 3.

EXPERIMENT 8. Solubility of wool and silk.

Take 2 g samples of wool and silk, wet them thoroughly in boiling water then squeeze out excess water. Shred them and add to separate 250 cm^3 beakers each containing 100 cm^3 concentrated hydrochloric acid. Stir for 15 minutes. The silk slowly dissolves; the wool, although swollen, does not dissolve because the disulphide cross-links are unaffected. Filter the silk solution through glass wool and add it slowly with stirring to an excess of cold water. Note the precipitation of finely divided silk fibroin.

TABLE 3   Differences Between Wool and Silk

	Wool	Silk
Formation of fibre	Growing (deposition)	Spinning (extrusion)
Form	Staple	Continuous filament
Surface	Scaly	Smooth
Sulphur-containing amino-acids	PRESENT	ABSENT
Covalent cross-links	PRESENT	ABSENT
Side-chain amide groups	PRESENT	ABSENT
Frequency of main amino-acids	None more than 12% of total	Over 90% derived from 4 amino-acids, about 75% from 2 amino acids
Arrangement of amino-acids	No major repeating unit	Long sequences of alternating glycine and alanine units believed present
Crystallinity	Very low	High
Chain conformation	α-helix	β-helix

Regenerated protein fibres

Most readers will be familiar with the stringy filamentary product that can be drawn from the surface of melted cheese after a few minutes heating. The viscosity increase that occurs on keeping cheese molten is connected with conversion of the protein casein from its natural globular or curled-up form into an extended form. This transformation is brought about in another, more controllable way in making fibres from casein extracted from milk, and also from the vegetable proteins extracted from ground-nuts, maize and soya beans. The protein is dissolved in dilute sodium hydroxide, then the solution is stored to 'ripen' it, i.e. to let its viscosity increase to the desired level. The polymer chains now uncoil and become more extended. Once the right viscosity is reached, the solution is spun into a coagulating bath containing dilute sulphuric acid and sodium sulphate.

If the fibres were now washed and dried, they would be strong when dry but very weak when wet. They are therefore treated with formaldehyde and an acidic catalyst such as aluminium sulphate. The formaldehyde links chains together (a process known chemically as cross-linking, and in this context as *hardening*) by reacting with amide groups in the protein:

$$
\begin{array}{c}
) \\
) \\
O{=}C \\
| \\
N{-}H
\end{array}
+ CH_2O +
\begin{array}{c}
) \\
) \\
C{=}O \\
| \\
H{-}N \\
) \\
)
\end{array}
\xrightarrow{\text{catalyst}}
\begin{array}{c}
) \\
) \\
O{=}C \\
| \\
N{-}CH_2{-}N \\
) \\
)
\end{array}
\begin{array}{c}
) \\
) \\
C{=}O \\
|
\end{array}
+ H_2O
$$

Regenerated protein fibres are nearer to wool in most of their properties than are any other man-made fibres—hardly surprising, since wool is a protein fibre of rather similar composition. For example, the moisture absorption and the heat developed on absorbing moisture are very similar for wool and regenerated protein fibre. There is, however, an important physical difference, since the man-made fibre lacks the scaly surface of the natural fibre and therefore does not felt. There is also an important chemical difference, for wool, as we have already seen, contains cross-links derived from the amino-acid cystine, whereas cystine is absent from the man-made fibres.

In spite of their resemblance to wool regenerated protein fibres have never been a major commercial success. Although they have been made commercially since the late 1930's, the amount produced is small relative to the regenerated cellulose fibres.

5 Fibres from synthetic polymers

5.1 Fibre manufacture and properties

Nylon-6,6

The word 'nylon', although chosen by Du Pont as a name for their new polyamide fibre, was not registered as a trade-mark. It has come to be a generic name for linear synthetic polyamides of different chemical structures made by many different manufacturers. '6,6' denotes that this particular nylon is derived from the 6-carbon diamine hexamethylenediamine and the 6-carbon acid adipic acid.

The first step in nylon-6,6 manufacture is to make a salt from hexamethylenediamine and adipic acid by mixing equimolar proportions of the base and acid each dissolved in water:

$$H_2N(CH_2)_6NH_2 + HOOC(CH_2)_4COOH \longrightarrow H_3\overset{+}{N}(CH_2)_6\overset{+}{N}H_3 \ \overset{-}{O}_2C(CH_2)_4 \overset{-}{C}O_2$$

hexamethylenediamine   adipic acid   6,6 salt

After treatment with activated carbon to remove impurities, the salt solution is heated until a pressure of several atmospheres is reached. Steam is bled off, and the temperature is raised until it reaches about 275 °C. The reason for developing a pressure is that if the water were removed at too low a temperature, solid polymer would separate and only re-melt with difficulty. At 275 °C the temperature is above the polymer melting point, so the pressure is allowed to fall to atmospheric and water removal continued until the desired molecular weight is reached.

$$n \ H_3\overset{+}{N}(CH_2)_6\overset{+}{N}H_3O_2C(CH_2)_4\overset{-}{C}O_2 \xrightarrow[n \ H_2O]{\text{loss of}} \Big[NH(CH_2)_6NHCO(CH_2)_4CO \Big]_n$$

6,6 salt   repeat unit of nylon-6,6

The principal end-groups are amino (—NH$_2$) and carboxyl (—COOH), but in order to help control the final molecular weight a small amount of a stabilizer, usually acetic acid, may be added to the salt before polymerization; this forms acetamido (CH$_3$CONH—) end-groups, which react less readily than amino with carboxyl groups and therefore limit the molecular weight.

Most nylon polymer is made in batches. As each batch reaches the required molecular weight it is extruded from the autoclave—which may hold several tonnes—by opening a valve at the foot and applying pressure inside the autoclave. The molten polymer is formed into a ribbon by falling on to a rotating water-cooled drum. When cool it passes through a machine which breaks the ribbon into small chips which are easier to store and transport. These can later be fed to the melting system of a spinning head. It is also common, however, to make the molten polymer continuously by feeding 6,6 salt to one end of a polymerization vessel and removing water so that the required molecular weight is reached at the other end of the vessel. The molten polymer then flows directly to the spinning head without ever solidifying.

Nylon-6,6 is melt-spun. The filaments rapidly begin to crystallize and to absorb moisture. The drop in glass transition caused by moisture absorption encourages crystallization at room temperature. If this occurs after the filaments have been collected on a bobbin the yarn sloughs and will not unwind evenly at the drawing stage. So the yarn is crystallized before it reaches the bobbin, usually by passing it through a short tube filled with steam. The subsequent drawing process does not require application of heat to the spun yarn, although heat is sometimes applied to assist drawing in making particularly strong or heavy denier products.

Nylon-6,6 readily gives very strong fibres with excellent recovery from low and moderate extensions, stiffer than wool but less stiff than cotton and polyester, and having excellent abrasion resistance. These properties make it particularly valuable for ropes, tyre cords, carpets, stockings, and filament apparel. It is popular in the U.K. for shirts, yet in the U.S.A. nylon shirts are seldom worn—a consequence of the relatively hydrophobic nature of the fibre, which makes a close-knit structure uncomfortable in many American climatic conditions but pleasant to wear under the less hot and humid British conditions.

EXPERIMENT 9. Preparation of nylon-6,6.

Dissolve 57.6 g purified adipic acid in 500 cm³ warm methanol, and 46.4 g purified hexamethylenediamine (CARE: corrosive) in 100 cm³ methanol. Cautiously mix the cold solutions, cool to room temperature, filter off the crystalline 6,6 salt and dry it in air. Place the 6,6 salt in a polymerization tube (Plate 1) fitted with a nitrogen inlet, flush the tube with nitrogen, and heat it gently in dimethyl phthalate vapour (b.p. 282 °C). Continue heating with a slow nitrogen stream; water is eliminated and is condensed and collected in the receiver. The product becomes steadily more viscous. After about 2 hours extrude it from the polymerization tube by turning off the heat supply to the vapour bath, removing the tube from the bath, nicking and knocking off the glass extension at the foot of the tube, and applying gentle nitrogen pressure to the tube. Alternatively wrap the tube in a cloth and leave it to cool, then remove the polymer by breaking the cold tube. Fibres may be drawn from the hot molten polymer.

Nylon-6

Many cyclic amides (or *lactams*) are readily converted into linear polymers. The most important polymer of this type is nylon-6, produced from caprolactam:

$$n \text{ O=C—NH} \rightleftharpoons [\text{—(CH}_2)_5\text{CONH—}]_n$$
$$(\quad)$$
$$(\text{CH}_2)_5$$

ε-caprolactam   repeat unit of nylon-6

Pure caprolactam is stable when molten, but polymerizes rapidly if an initiator is present. For fibre production, water is a common initiator. It operates by opening the ring to form ε-aminocaproic acid; further lactam then adds on to the amino-group, and the addition process continues:

$$\text{O=C—NH} + \text{H}_2\text{O} \rightleftharpoons \text{HOOC(CH}_2)_5\text{NH}_2$$
$$(\quad)$$
$$(\text{CH}_2)_5$$

caprolactam   adds ↓ caprolactam

$$\text{HOOC(CH}_2)_5\text{NHCO(CH}_2)_5\text{NH}_2$$

adds ↓ caprolactam

$$\text{HOOC(CH}_2)_5\text{NHCO(CH}_2)_5\text{NHCO(CH}_2)_5\text{NH}_2$$

adds ↓ caprolactam

progressively higher molecular weight

This stage is carried out at 220–240 °C under pressure to prevent loss of water and lactam. The product is however of a lower molecular weight (usually 8000–14 000) than is ultimately required. So far polymerization has occurred by an addition mechanism; now the temperature is raised and the pressure is reduced to below atmospheric to encourage removal of water and increase of molecular weight by a condensation mechanism:

$$\text{H}_2\text{N[(CH}_2)_5\text{CONH]}_m(\text{CH}_2)_5\text{COOH} + \text{H}_2\text{N[(CH}_2)_5\text{CONH]}_n(\text{CH}_2)_5\text{COOH}$$
$$\downarrow -\text{H}_2\text{O}$$
$$\text{H}_2\text{N[(CH}_2)_5\text{CONH]}_{(m+n+1)}(\text{CH}_2)_5\text{COOH}$$

Finally the polymer is heated under nitrogen at about 270 °C to attain equilibrium with the water and cyclic monomer present, at which point it has a molecular weight above 20 000. To help control the final molecular weight a stabilizer—an amine or carboxylic acid—is usually present from the beginning of the polymerization; it acts in the same way as acetic acid in nylon-6,6 polymerization, by reacting with the end of a polymer chain and so blocking it against further condensation with other polymer chains.

Some unchanged caprolactam remains in the polymer at equilibrium. It is removed from the polymer chips by washing them with hot demineralized water until the final caprolactam content is about 0.5%, then the polymer is dried to a controlled moisture content before spinning.

Because the melting point of nylon-6 is only about 215 °C, a lower spinning temperature can be used than for nylon-6,6. Otherwise the processing and end-uses of the two fibres are similar.

EXPERIMENT 10. Preparation of nylon-6.

Into a polymerization tube (Plate 1) equipped with a side-arm and capillary nitrogen inlet place 50 g ε-caprolactam and 2.0 g ε-aminocaproic acid in 2 cm³ water. Pass nitrogen slowly through the tube and heat in a dimethyl phthalate vapour bath, gently at first then more vigorously, for 4 hours. Extrude or isolate as in Experiment 9.

Polyester fibres

Although polyester fibres of several different chemical structures are manufactured commercially, much the most important are those made from poly(ethylene terephthalate). The polymer is made by condensing ethylene glycol with either terephthalic acid or its dimethyl ester. There are usually two distinct stages in its manufacture. The first is reaction of an excess of glycol with the acid or ester. Elimination of water or methanol forms a mixture of glycol esters of terephthalic acid, represented in the following scheme by the predominant species, bishydroxyethyl terephthalate. Then follows elimination of glycol from the hydroxyethyl ester to form a high polymer.

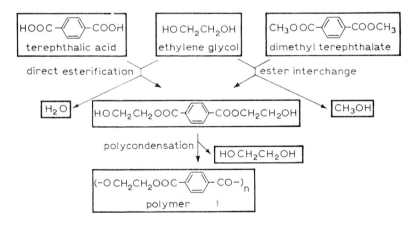

Direct esterification is self-catalysed by the carboxylic acid groups, but ester interchange and polycondensation—formally the same type of reaction—both need catalysts. The effective catalysts are mostly alkoxides of metals such as calcium, manganese, cobalt, magnesium, or zinc for the first stage, and such as antimony or germanium for the second.

An essential feature of the polycondensation stage is the use of very low pressure to remove the glycol produced from the system; because the forward and back reactions are very similar in form and in rate, displacement of the equilibrium to the right requires the driving force of a very low

$\sim$OOC—⬡—COOCH$_2$CH$_2$CH $\sim$OOC—⬡—CO

$+$ $\rightleftharpoons$

$\sim$OOC—⬡—COOCH$_2$CH$_2$CH $\sim$OOC—⬡—CO

$$\begin{array}{cc} & O \\ & | \\ & CH_2 \quad CH_2OH \\ & | \quad + \quad | \\ & CH_2 \quad CH_2OH \\ & | \\ & O \end{array}$$

two end groups   new chain link   glycol

partial pressure of glycol in the polymerization vessel. Quite often the dissolved glycol is not in equilibrium with that in the vapour space. The reaction rate is then partly controlled by the rate of diffusion of glycol through the molten polymer to the surface.

Like the nylons, poly(ethylene terephthalate) can also be made continuously instead of batchwise. The maximum advantage is obtained from continuous polymerization if the product is fed directly from the final polymerization vessel to the melt-spinning heads so that the processes of casting, chipping, blending, and drying necessarily associated with batchwise operation are completely eliminated.

Poly(ethylene terephthalate) melts at about 260 °C and is melt-spun at temperatures of 280–300 °C; above 300 °C it decomposes too quickly. The temperatures resemble those for nylon-6,6, yet the structural differences lead to some significant differences in processing. For example:

(i) Molten polyester is much more readily degraded by hydrolysis than molten polyamide, because in the reversible reactions

$$RCOOR^1 + H_2O \underset{k_2}{\overset{k_1}{\rightleftharpoons}} RCOOH + R^1OH$$

$$RCONHR^1 + H_2O \underset{k_4}{\overset{k_3}{\rightleftharpoons}} RCOOH + R^1NH_2$$

the equilibrium constant k_1/k_2 is very much greater than k_3/k_4. This means that whereas the nylons can be melt-spun with water present, polyester must be rigorously dried before it is melted otherwise a serious fall in molecular weight will take place.

(ii) The polymer chain is relatively inflexible, so the melt viscosity is much higher than in the polyamides. One result is a higher spinning pressure.

(iii) Its inflexibility and low moisture absorption keep the glass-rubber transition of the polyester well above room temperature, so it fails to crystallize during spinning. Moreover, drawing in the glassy state is an unstable process. The yarn must therefore be heated to a temperature above the transition temperature for stable drawing.

Poly(ethylene terephthalate), like nylon-6,6, readily gives very strong fibres with excellent recovery from low extensions, but the fibre is stiffer and less hydrophilic. The stiffness makes it suitable for blending in staple form with cotton, which provides the moisture absorbing capacity to make the fabric comfortable in such uses as shirts. Bulked filament yarns for dresses, polyester/wool blends for trousers which provide outstanding crease retention, and curtain nets which utilize the high stability when exposed to light, are other important uses.

EXPERIMENT 11. Preparation of poly(ethylene terephthalate).

Into a glass polymerization tube (Plate 1), about 1.25 inches diameter × 10 inches length, fitted with a side-arm and a tube inserted through the neck and drawn out into a fine capillary at the foot, place 50 g dimethyl terephthalate, 40 g dry ethylene glycol, 0.05 g calcium acetate dihydrate, and 0.06 g antimony trioxide. Melt the reactants by immersing the lower half of the tube in the vapours of boiling ethylene glycol; pass a very slow stream of nitrogen through the capillary, and collect the methanol that distils over. After 1 hour raise the glycol vapour level to heat as much of the tube as possible and distil off methanol for a further hour. Dispose of the methanol.

Introduce 0.05 g phosphorous acid in a little glycol into the tube, then commence heating it in dimethyl phthalate vapour, gently at first to distil off excess glycol, then with complete immersion. After 10 minutes apply vacuum slowly to the distillate receiver, so that the pressure reaches 0.1 mm in about 15 minutes. Polymerize at as low a pressure as possible for 2 hours, then release the vacuum. Extrude or isolate as in Experiment 9.

Acrylic fibres

The chief component of the polymers from which acrylic fibres are spun is acrylonitrile, a cheap and simple unsaturated compound that readily polymerizes to polyacrylonitrile:

$$n \; CH_2{=}CH \longrightarrow (-CH_2{-}CH{-})_n$$

CN	CN
acrylonitrile	polyacrylonitrile

Polymerization of acrylonitrile is initiated by a free radical—a molecular fragment containing an unpaired electron—which adds to one of the ethylenic carbon atoms. This addition produces an unpaired electron on the other ethylenic carbon atom, which can then add to a further molecule of acrylonitrile. Such addition continues as a chain reaction so that by repeated addition of molecules of acrylonitrile a very large linear polymer molecule builds up. Eventually the chain is terminated by a meeting of the growing ends of two polymer molecules. In the following scheme dots indicate unpaired electrons.

$$R\text{—}R \longrightarrow 2R.$$

$$R. + CH_2{=}\overset{\overset{\displaystyle CN}{|}}{C}H \longrightarrow R\text{—}CH_2\text{—}\overset{\overset{\displaystyle CN}{|}}{C}H.$$ $\Big\}$ initiation

$$R\text{—}CH_2\text{—}\overset{\overset{\displaystyle CN}{|}}{C}H. + CH_2{=}\overset{\overset{\displaystyle CN}{|}}{C}H \longrightarrow R\text{—}CH_2\text{—}\overset{\overset{\displaystyle CN}{|}}{C}H\text{—}CH_2\text{—}\overset{\overset{\displaystyle CN}{|}}{C}H.$$

$$\text{\tiny\textasciitilde\textasciitilde\textasciitilde}CH_2\text{—}\overset{\overset{\displaystyle CN}{|}}{C}H. + CH_2{=}\overset{\overset{\displaystyle CN}{|}}{C}H \longrightarrow \text{\tiny\textasciitilde\textasciitilde\textasciitilde}CH_2\text{—}\overset{\overset{\displaystyle CN}{|}}{C}H\text{—}CH_2\text{—}\overset{\overset{\displaystyle CN}{|}}{C}H.$$ $\Big\}$ propagation

$$2\,\text{\tiny\textasciitilde\textasciitilde\textasciitilde}CH_2\text{—}\overset{\overset{\displaystyle CN}{|}}{C}H. \longrightarrow \text{\tiny\textasciitilde\textasciitilde\textasciitilde}CH_2\text{—}\overset{\overset{\displaystyle CN}{|}}{C}H\text{—}\overset{\overset{\displaystyle CN}{|}}{C}H\text{—}CH_2$$

$$2\,\text{\tiny\textasciitilde\textasciitilde\textasciitilde}CH_2\text{—}\overset{\overset{\displaystyle CN}{|}}{C}H. \longrightarrow \text{\tiny\textasciitilde\textasciitilde\textasciitilde}CH{=}\overset{\overset{\displaystyle CN}{|}}{C}H + \overset{\overset{\displaystyle CN}{|}}{C}H_2\text{—}CH_2\text{\tiny\textasciitilde\textasciitilde\textasciitilde}$$ $\Big\}$ termination

The radical source R—R is a compound, often a peroxidic material such as a persulphate, that readily decomposes to give two free radicals R. each of which acts as initiator of a polymerization chain. Much of the polyacrylonitrile required for fibre production is made by polymerizing acrylonitrile dissolved along with the radical source in water. Since polyacrylonitrile is insoluble in water it precipitates as it is formed.

Other unsaturated compounds, such as methyl acrylate

$$CH_2{=}CH\text{—}COOCH_3$$

will react like acrylonitrile if a mixture of the two is polymerized, so that the product contains units derived from more than one monomer. Acrylic fibres are made from such co-polymers in which acrylonitrile provides at least 85% by weight of the total. There is another, related, class known as *modacrylic fibres* that contain much less acrylonitrile, but they are of less importance.

Polyacrylonitrile will not melt without decomposition, so the fibres are made by dry- or wet-spinning from solution. Solvents for polyacrylonitrile are not easy to find. Indeed, it was the discovery of simple solvents and the development of processes for their production that permitted commercial production of acrylic fibres to begin in 1950. These first solvents were N,N-dimethyl formamide and N,N-dimethyl acetamide, both very powerful, highly polar solvents. Some inorganic solvents are also in commercial use nowadays; they include strong aqueous calcium or sodium thiocyanate, 60% zinc chloride, and 70% nitric acid. Because of their involatility they are used for wet-spinning but not for dry-spinning.

Acrylic fibres are usually considered to exhibit some molecular order, but not crystallinity, and are less strong than the major polyester and polyamide fibres. The lower strength makes them unsuitable for most industrial end-uses, but is perfectly satisfactory for domestic purposes such as furnishings, and for knitting yarns, where their soft, pleasant handle makes them popular. Carpets, knitted outerwear, pile fabrics for furnishings, and blankets are among the chief uses.

EXPERIMENT 12. Polyacrylonitrile.

Pass about 100 g of commercial acrylonitrile (CARE: toxic vapour) down a column of dry silica gel one inch wide and two feet long to remove the polymerization inhibitor. In a 250 cm³ flask equipped with a stirrer, reflux condenser, thermometer pocket, and nitrogen inlet boil 100 cm³ of distilled water for 5 minutes, then place the flask in a thermostat bath at 30 °C and allow the contents to cool to 30 °C while passing a gentle stream of nitrogen through the system. Add 1 g sodium lauryl sulphate, 50 g purified acrylonitrile, 0.1 g potassium persulphate, and 0.05 g sodium hydrogen sulphite, stirring well throughout. Note a rise in temperature due to exothermic polymerization, and development of a milky appearance as the insoluble polymer separates as an emulsion. Continue to stir at 30 °C for at least 3 hours, then pour the product into 500 cm³ of 10% sodium sulphate solution. The emulsion coagulates. Filter the product off, wash well with water, and dry in a vacuum desiccator. Measure the ratio of the viscosity of a 1% solution in N,N-dimethyl formamide at 25 °C to that of the pure dimethyl formamide. A ratio of 1.4 indicates a molecular weight of about 20 000 and of 2.4 about 100 000.

Dissolve 20 parts of the polymer in 80 parts of dimethyl formamide by adding the finely divided polymer to the cold solvent then heating with agitation at 100 °C. This solution may be used to wet-spin fibres, using water or aqueous dimethyl formamide as coagulant and the techniques described on p. 16.

Polypropylene fibres

The polyalkenes, unlike the polymers already considered, are more important as moulding plastics than as fibres. Polyethylene, which at best melts no higher than 140 °C, is very limited in its fibre applications, but polypropylene, which melts at about 167 °C, is more useful.

Propylene, $CH_2=CH—CH_3$, is a petroleum cracking product, plentiful and readily separated from other products and therefore very cheap. Because it contains a methyl group attached to one of the double-bonded carbon atoms it cannot be turned into a high polymer by a free radical polymerization as can ethylene, $CH_2=CH_2$. Even if it could the product would probably be a non-crystalline gum because the methyl groups on alternate carbon atoms can occupy either of two positions relative to each other.

$$
\begin{array}{cccc}
& & \overset{3}{CH_3} & \\
H & H & | & H \\
| & | & | & | \\
-CH_2-C-CH_2-C-CH_2-C-CH_2-C- \\
| & | & | & | \\
CH_3 & CH_3 & H & CH_3 \\
1 & 2 & & 4
\end{array}
$$

In the planar formula above, the methyl group marked 2 is in the same position relative to the chain as that marked 1; that marked 3 is on the opposite side from 2; that marked 4 on the opposite side from 3. There is no reason, during a free radical polymerization, why the methyl groups should adopt one position or the other so they end up randomly oriented relative to the chain and each other. The term *atactic* is applied to a polymer of this type. Because of the random orientation of the substituents it is non-crystalline.

In 1954 the Italian G. Natta, in extending some discoveries of the German K. Ziegler, found that propylene could be polymerized by a new type of catalyst to a crystalline high polymer. Investigation showed that the methyl groups were all on the same side of the plane of the main chain.

$$
\begin{array}{cccc}
H & H & H & H \\
| & | & | & | \\
-CH_2-C-CH_2-C-CH_2-C-CH_2-C- \\
| & | & | & | \\
CH_3 & CH_3 & CH_3 & CH_3
\end{array}
$$

Because of this regularity the polymer crystallizes. This kind of ordered polymerization is known as *stereo-regular* polymerization, and the polymer with this particular regularity is said to be *isotactic*. There are other types of regularity, for instance where the methyl groups are on alternate sides of the chain, but the isotactic polymer is the commercially important one.

The special catalysts used for this isotactic polymerization are often complexes formed from titanium trichloride, $TiCl_3$, and an aluminium trialkyl AlR_3 or aluminium dialkyl chloride AlR_2Cl. The catalyst is used as a suspension in an alkane fraction, and converts propylene, introduced under pressure, into a suspension of solid polymer in the alkane. The solid polymer is isolated and purified as far as possible by washing out catalyst residues.

Special steps must be taken to overcome two deficiencies of polypropylene: it is rather readily oxidized if exposed to heat or light, and it contains no sites for dyes to attach themselves to. Polypropylene fibre always contains antioxidants to suppress the oxidation—quite often as many as three different antioxidants, one to absorb ultra-violet radiation (which would start oxidation chains in the polymer) and convert it into harmless longer wave-lengths; one to stop oxidation chains by reacting with the free radicals and removing them; and one to stop oxidation chains by reacting with hydroperoxides before they can decompose to radicals.

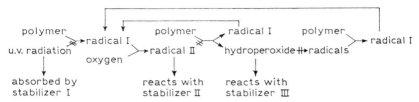

How three different kinds of stabilizer can protect polypropylene against oxidation.

Dyeability does not matter in all uses—not in ropes, for example—but where it does one of two methods is adopted. One is to add coloured pigments to the polymer and sell coloured fibre. The other is to add a 'dye site'—a material that will complex or react with dyes—and sell dyeable fibre.

Polypropylene is all melt spun, but usually at a temperature far above its melting point—sometimes even above 300 °C. The reason for the high temperature is two-fold. First, the molecular weight of the polymer is much higher than that of the more polar melt-spun fibres. This makes the melt exceptionally viscous even though the chain contains quite flexible links. Second, the molecular weight of the polymer has a much wider spread than usual. This results in peculiar behaviour on extrusion, which is minimized by working at a higher extrusion temperature.

Elastomeric fibres

All the fibre structures discussed so far have quite low extensibilities—the fibres break on extension long before they reach double their initial length. Usually, too, they exhibit a yield point, and if stretched beyond this yield point show poor recovery. However, fibres do exist that can be extended to far more than double the initial length, and exhibit almost complete recovery from very high extensions. Elastic made from rubber is the best known of such materials; they are said to be *elastomeric*.

The chemical structure needed for such behaviour is well defined. The fibre must contain two kinds of group alternating within each polymer molecule:
(1) Relatively long segments of chain of a very flexible nature—i.e. in the rubbery state and possessing little or no crystallinity. These are known as 'soft' segments.
(2) Relatively short segments of chain providing strong inter-chain forces. These may be covalent cross-links (as in vulcanized rubber) or rigid, usually crystalline and polar, units known as 'hard' segments.

Although it is possible to produce fibres of this type from many synthetic rubbers, the most successful synthetic elastomeric fibres possess a

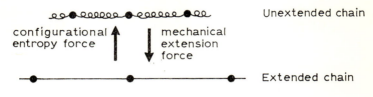

configurational entropy force ↑   ↓ mechanical extension force   Unextended chain

Extended chain

● = Hard segments   ℓℓℓ = Soft segments

FIG. 5.1   Schematic representation of an elastomeric polymer chain.

structure classified as 'Spandex', based on aliphatic polyester or polyether soft segments plus hard segments built up through isocyanate addition reactions. One well-known Spandex fibre, Lycra, is built up in rather a complicated way from soft segments of a polyether made by polymerizing tetrahydrofuran to a relatively low molecular weight. The soft segment is enlarged and 'capped' with isocyanate groups by reaction with an excess of di-isocyanate:

$$CH_2—CH_2$$
$$CH_2 \quad CH_2 \quad \xrightarrow{} HO(CH_2CH_2CH_2CH_2O)_{n-1} CH_2CH_2CH_2CH_2OH$$
$$\diagdown O \diagup$$

tetrahydrofuran

$$\downarrow OCN\,R\,NCO$$

$$[O\,CN\,R\,NH\,COO(CH_2CH_2CH_2CH_2O)_n\,CO\,NH]_2R$$

This whole unit, plus any excess unreacted di-isocyanate, is now caused to react with a diamine such as hydrazine; to ensure that the molecular weight does not become unmanageably high a little mono-amine is also included. This final stage is carried out in a solvent such as dimethyl formamide. The molecule is thus built up through reactions such as

$$OCN \overset{}{\underset{\text{soft}}{\sim\sim}} NCO + H_2NNH_2 + OCN \overset{}{\underset{\text{soft}}{\sim\sim}} NCO$$

$$\downarrow$$

$$OCN \overset{}{\underset{\text{soft}}{\sim\sim}} \underset{\text{hard}}{NHCONHNHCONH} \overset{}{\underset{\text{soft}}{\sim\sim}} NCO$$

to a very high molecular weight.

The product of the series of chemical reactions is a solution of the polymer, in which hard and soft segments now alternate, in the solvent used for the final stage. This particular fibre is then dry-spun.

The structural reasons for the elastic behaviour of such fibres are easily understood. The soft segments are easily deformed, with no high energy barriers to rotation about bonds, so low stresses produce high extensions (i.e. the modulus is low). The hard segments are not deformed, but act as tie units which hold the ends of the soft segments together. Once the stress is removed, the soft segments tend to return to their natural average end-to-end length and so the fibre returns to its original length.

Carbon fibres

Some of the first successful work on man-made fibre production was due to the need of the electric light industry for a cheap conducting filament for light bulbs. Swan made filaments by spinning nitrocellulose, and later viscose, then pyrolysing the filaments to convert them into carbon. The modern use of carbon fibres is quite different, and depends on their very high stiffness, which weight for weight is several times greater than that of steel. Carbon fibres are ideal for stiffening thermosetting resins to form very rigid structures whose lightness, compared with alternative materials, suits them for such uses as turbine blades in jet engines.

Although some carbon fibre is still made by pyrolysing cellulosic fibres, very high modulus carbon fibres are now produced from highly oriented acrylic fibres which are pyrolysed and oxidized very slowly under carefully controlled conditions to convert them into carbon. Acrylic fibre is particularly effective because pyrolysis converts polyacrylonitrile into an unusual polycyclic polymer, already part way to the graphite carbon structure:

polyacrylonitrile

a polycyclic intermediate

a section of the planar graphite structure

The structural reason for the stiffness of carbon fibres is obvious: since there are no flexible chains present, there is an immediate high resistance to extension. The breaking load is high, but the extension required to break the fibre is very low compared with textile fibres. The fibre is therefore brittle; it cannot be bent far without breaking. Even quite low bending angles produce too high an extension on the outside of the bend. It is therefore useless as a conventional textile fibre, but as a component of composite materials for engineering use it has enormous potential. At present there is one serious drawback: its cost. Although the acrylic fibre from which it is made is quite cheap, the pyrolysis process must be carried out very slowly to make a useful product—so slowly that the process is a costly one and the product correspondingly expensive.

5.2   Intermediates for synthetic fibres

In discussing the chemical structure and manufacture of the polymers used in synthetic fibres, the chemical intermediates were taken for granted. But millions of tons a year of synthetic fibres require millions of tons of intermediates—an enormous chemical industry in itself. It may be instructive to look at some commercial syntheses of the key intermediates for the three major classes of synthetic fibre—nylons (i.e., polyamides), polyesters, and acrylics—and see how the rise of the synthetic fibre industry has affected the course of industrial organic chemistry, and vice versa.

Nylon-6,6

The development of nylon-6,6 provides a good example. The two intermediates concerned are hexamethylenediamine, $H_2N(CH_2)_6NH_2$, and adipic acid, $HOOC(CH_2)_4COOH$. One of the early decisions the development team at du Pont had to make was which nylon to choose out of several, all of which gave good fibre properties. Nylon-6,6 is higher melting than most of the others, but it is also less colour-stable and undergoes side-reactions in the melt more readily. Another, more stable, nylon, such as 6,10, might well have been chosen but for the predicted cost of the intermediates. The 10-carbon acid (sebacic acid) and other candidates were unlikely to be available from any cheap starting material, whereas adipic acid and hexamethylenediamine, then of no commercial importance, were potentially available from the cheap coal-tar primary, phenol. So nylon-6,6 was chosen, and a manufacturing process based on phenol was developed:

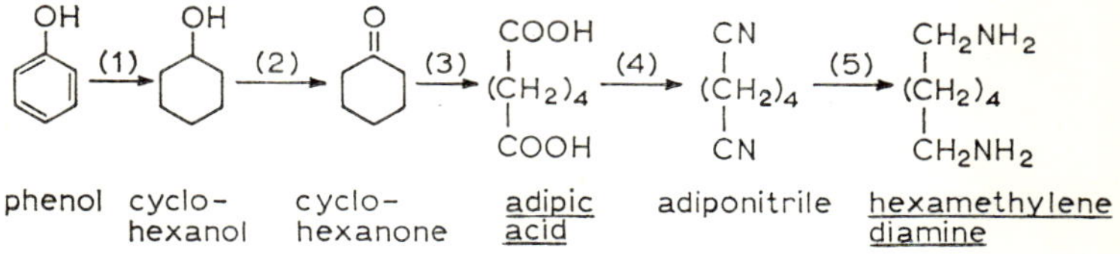

phenol   cyclo-   cyclo-   adipic   adiponitrile   hexamethylene
         hexanol   hexanone   acid   diamine

(1) Catalytic hydrogenation
(2) Catalytic air oxidation
(3) Catalytic air oxidation or nitric acid oxidation
(4) Vapour phase reaction with ammonia over a dehydrating catalyst, boron phosphate
(5) Catalytic hydrogenation, cobalt catalyst

Phenol is cheap, but benzene is cheaper. So the next development was based on benzene:

benzene   cyclohexane   cyclohexanol + cyclohexanone   adipic
                                   'KA'(ketone + alcohol)   acid

(1) Catalytic hydrogenation, nickel catalyst
(2) Catalytic air oxidation, cobalt salt catalyst
(3) Oxidation by nitric acid, Cu/V catalyst

This has been the major process for adipic acid for many years, though recent improvements have now overcome the yield problem once associated with air oxidation as a means of producing adipic acid from cyclohexane or its partial oxidation products.

Although adipic acid is cheap by these routes hexamethylenediamine requires the two further, less efficient, stages (4) and (5) shown in the first scheme so more attention has been devoted to new routes to the diamine, three of which are outlined below:

I $CH_2{=}CH{-}CH{=}CH_2$ $\xrightarrow{\text{(1)}}$ mixed
                butadiene   dichlorobutenes
                                                 $\downarrow$ (2)

    $NC{-}CH_2{-}CH{=}CH{-}CH_2{-}CN$ $\xleftarrow{\text{(3)}}$ mixed
           1,4-dicyanobut-2-ene   dicyanobutenes
               $\downarrow$ (4)

              $NC{-}(CH_2)_4{-}CN$ $\xrightarrow{\text{(5)}}$ $H_2N{-}(CH_2)_6{-}NH_2$
               adiponitrile   hexamethylenediamine

             (1) HCl, O_2, $CuCl_2$ catalyst   (4) hydrogenation
             (2) HCN, Cu powder   (5) hydrogenation
             (3) isomerization

II   Cyclohexane $\xrightarrow{\text{(1)}}$ $HOOC(CH_2)_4COOH + HO(CH_2)_5COOH$
                                           adipic acid   ε-hydroxycaproic
                                                                            acid
                           $\downarrow$ (2)

    $HO(CH_2)_6OH$ $\xleftarrow{\text{(3)}}$ mixed esters
    hexane-1,6-diol
        $\downarrow$ (4)   (1) Oxidation, Co^{2+} catalyst
                        (2) Esterification with hexane-1,6-diol
    $H_2N(CH_2)_6NH_2$ (3) Hydrogenation
    hexamethylenediamine   (4) NH_3, Ni catalyst

III $2\,CH_2{=}CH{-}CN$ $\xrightarrow{\text{(1)}}$ $NC(CH_2)_4CN$ $\xrightarrow{\text{(2)}}$ $H_2N(CH_2)_6NH_2$
        acrylonitrile   adiponitrile   hexamethylenediamine

            (1) Hydrodimerization   (2) Hydrogenation

This last process is a particularly novel one, which depends on a reaction at the cathode of an electrolytic cell known as 'reductive hydrodimerization'. It is noteworthy that this route would not have been competitive with the others but for the development of the cheap 'ammoxidation' route to acrylonitrile discussed later.

Nylon-6

ε-Caprolactam, the source of nylon-6, is mostly made from cyclohexanone, which as we have just seen is available from benzene or phenol, by a route which is virtually a scaled-up textbook laboratory method. It employed the Beckmann isomerization of oximes to amides, thus:

cyclohexanone   cyclohexanone   ε-caprolactam
                         oxime

One of the newer routes is particularly interesting as probably the largest scale use of ultra-violet radiation in chemical synthesis. This process, developed in Japan, converts cyclohexane directly into cyclohexanone oxime by reacting it with nitrosyl chloride under u.-v. radiation, which splits the NO—Cl bond and thus catalyses the reaction:

cyclohexane   cyclohexanone
                           .oxime

The oxime is then isomerized to the lactam as above.

Polyester intermediates

The key intermediates in polyester fibre manufacture are terephthalic acid and its ester dimethyl terephthalate, either of which can be used, if pure enough, for making polymer. When the fibre was discovered both were little known compounds; now each is produced in hundreds of thousands of tons a year.

The main starting material is para-xylene, which must be separated from the mixed 8-carbon aromatic petroleum fraction. Ortho-xylene and ethylbenzene are readily separated from the other two isomers by fractional distillation, but meta-xylene and para-xylene boil only 0.7 °C apart so they are usually separated by a freezing process. Pure para-xylene freezes at 13.26 °C, so para-rich product crystallizes first and is separated from meta-rich product. This process can be made to produce very pure para-xylene, but there is little use for the meta-xylene which was the major constituent of the original mixed xylene. So the meta-xylene is isomerized by passing the vapour over a silica-alumina or platinum catalyst at 450 °C to form a mixture of xylenes, which is re-cycled to the separation process:

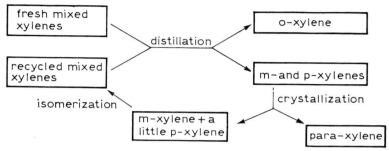

This process was developed for polyester production along with an oxidation process for converting p-xylene into terephthalic acid using dilute nitric acid under pressure at a temperature of 180–200 °C. Because the terephthalic acid was rather impure and very difficult to purify it was then esterified with methanol and sulphuric acid—again under pressure—to dimethyl terephthalate, which can be purified by successive distillation and re-crystallization.

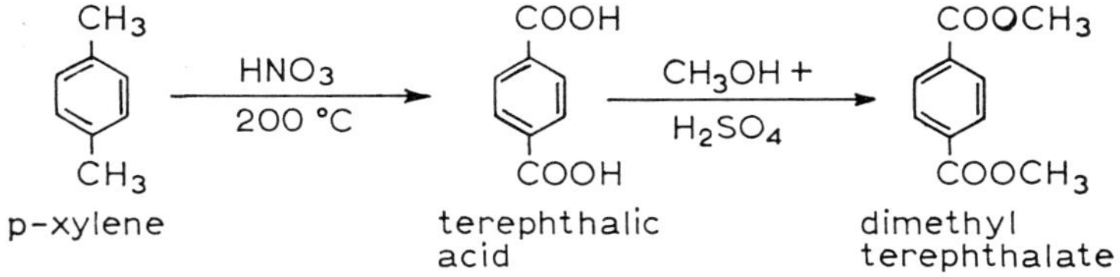

CH$_3$ ⬡ CH$_3$ $\xrightarrow[\text{200 °C}]{\text{HNO}_3}$ COOH ⬡ COOH $\xrightarrow[\text{H}_2\text{SO}_4]{\text{CH}_3\text{OH} +}$ COOCH$_3$ ⬡ COOCH$_3$

p-xylene   terephthalic acid   dimethyl terephthalate

Once this process was established there was a world-wide search for better routes, and the successful developments were all quite new. Some were based on p-xylene and the use of air as the oxidizing agent. The difficulty is not to oxidize p-xylene with air but to oxidize the intermediate product, p-toluic acid. One process is based on the discovery that although oxidation in acetic acid under pressure using the well-known oxidation catalysts, Co^{2+} and Mn^{2+} ions, fails to produce much terephthalic acid, the further presence of bromide anions catalyses formation of terephthalic acid in high yield.

EXPERIMENT 13. Oxidation of p-xylene to terephthalic acid.

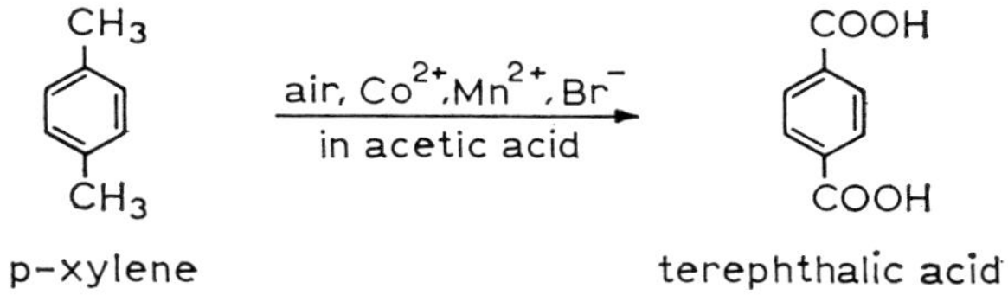

CH$_3$ ⬡ CH$_3$ $\xrightarrow[\text{in acetic acid}]{\text{air, Co}^{2+},\text{Mn}^{2+},\text{Br}^-}$ COOH ⬡ COOH

p-xylene   terephthalic acid

D

To 200 cm³ propionic acid in an oxidation flask (Fig. 5.2) add 40 g p-xylene, 0.2 g cobalt acetate tetrahydrate, 0.1 g manganese acetate tetrahydrate, and 0.2 g sodium bromide. Heat the solution to gentle reflux, then pass oxygen into the flask, with rapid agitation, at about 10 litres/hour. The reaction mixture changes colour then solid terephthalic acid begins to precipitate. Continue oxidizing for 2 hours, then stop passing oxygen and filter the hot solution through a sintered glass funnel. Wash isolated terephthalic acid with acetic acid then methanol and dry and weigh it. Notice that when heated on a spatula above a flame it does not melt, but sublimes away.

In industrial practice this reaction is carried out in acetic acid under pressure and air is used instead of oxygen. In the laboratory propionic acid must be used to obtain a sufficiently high reaction temperature. Two hours is insufficient for maximum yield of terephthalic acid in propionic acid; longer reaction times will produce more.

Another oxidation process avoids the difficulty by oxidizing methyl p-toluate, and is operated as a four-stage synthesis as shown:

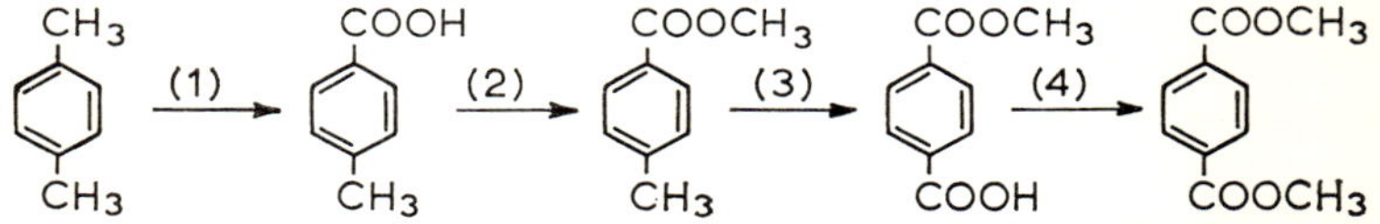

| p-xylene | p-toluic acid | methyl p-toluate | monomethyl terephthalate | dimethyl terephthalate |

Notice that these four stages comprise two oxidations and two esterifications. This process can also be operated so that the oxidations (1) and (3) are carried out in one vessel, and the esterifications (2) and (4) in one vessel, thus reducing the total number of vessels required.

Two other processes, closely related to each other, were developed by the German firm of Henkel but have only been used commercially in Japan. They depend on solid-phase isomerization or disproportionation of carboxylate groups:

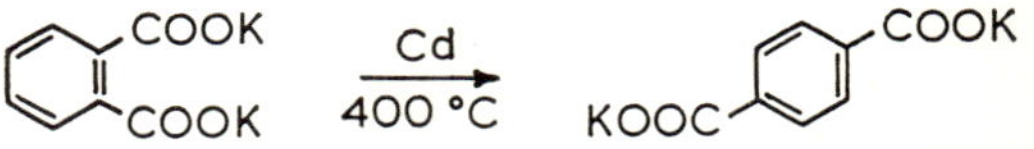

potassium phthalate → potassium terephthalate

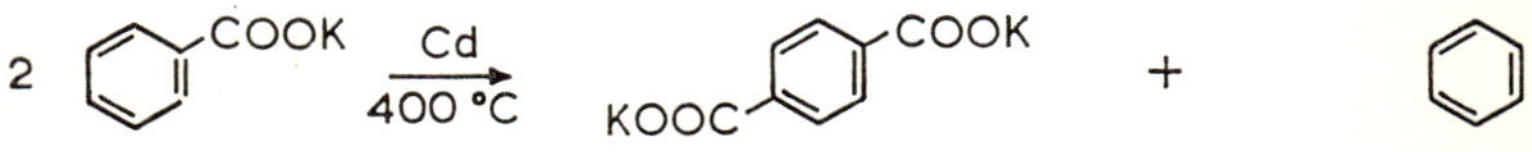

potassium benzoate → potassium terephthalate + benzene

The upper route is based on phthalic anhydride, available from oxidation of naphthalene or o-xylene, and the lower route on benzoic acid,

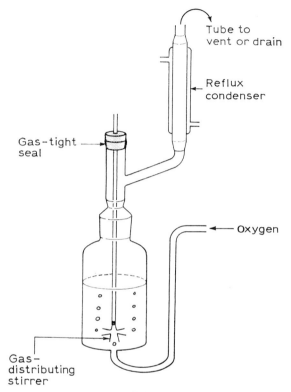

Tube to
vent or drain

Reflux
condenser

Gas-tight
seal

Oxygen

Gas-
distributing
stirrer

FIG. 5.2   Oxidation flask.

available from oxidation of toluene. In both cases the economics depend
on efficient potassium recycle, since potassium salts are rather expensive.
These processes are particularly interesting for the novelty of the chemical
approach and the use of high temperature solid-phase conditions.

The newer terephthalic acid processes give a purer product, and methods
of further purifying the acid have also been developed. Terephthalic acid,
although very insoluble under normal conditions, can be recrystallized
from water at a high temperature and pressure, and some further oxidation
or reduction of impurities carried out while it is in solution. These new
processes have made it possible to manufacture terephthalic acid pure
enough for polyester manufacture without any need to go through the
dimethyl ester.

The other important polyester intermediate is ethylene glycol, which
is made from the petroleum cracking product ethylene. Unlike most fibre
intermediates it was already important before development of the fibre—

it is the main constituent of anti-freeze. During recent years there has been a change from a synthesis based on ethylene chlorhydrin to direct oxidation of ethylene over a silver catalyst:

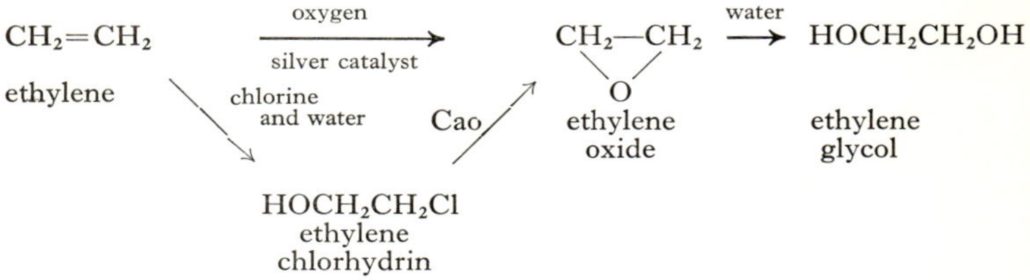

$$CH_2{=}CH_2 \xrightarrow[\text{silver catalyst}]{\text{oxygen}} CH_2{-}CH_2 \xrightarrow{\text{water}} HOCH_2CH_2OH$$

ethylene   chlorine and water   Cao   O ethylene oxide   ethylene glycol

HOCH$_2$CH$_2$Cl
ethylene
chlorhydrin

Acrylonitrile

The key intermediate for the remaining major group of synthetic fibres, the acrylics, again owes its importance to the use in fibres. In the earliest days of acrylic fibre development, acrylonitrile was made by base-catalysed addition of hydrogen cyanide to ethylene oxide then dehydration of the product:

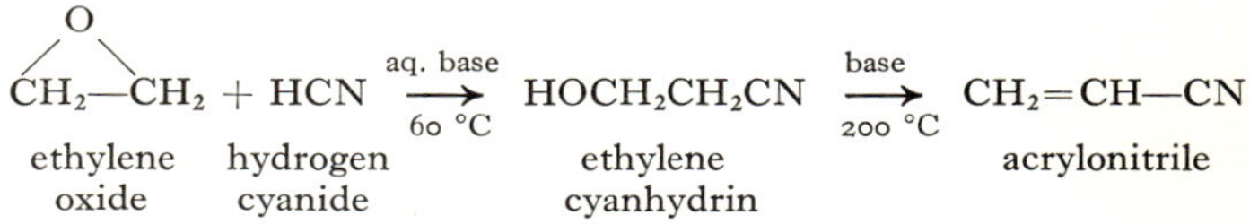

$$CH_2{-}CH_2 + HCN \xrightarrow[\text{60 °C}]{\text{aq. base}} HOCH_2CH_2CN \xrightarrow[\text{200 °C}]{\text{base}} CH_2{=}CH{-}CN$$

ethylene oxide   hydrogen cyanide   ethylene cyanhydrin   acrylonitrile

This two-stage synthesis was soon replaced by a single-stage synthesis by catalytic addition of hydrogen cyanide to acetylene. The catalyst in this case is copper (I) chloride, Cu$_2$Cl$_2$, and the sequence of reactions is:

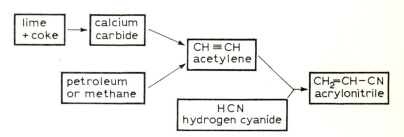

This process in turn has now been very largely supplanted by processes based on the petroleum cracking product, propylene, which is much cheaper than acetylene or hydrogen cyanide. The main process is known as 'ammoxidation', because it involves passing a mixture of ammonia and oxygen with acrylonitrile and steam over a heavy metal oxide catalyst such as bismuth phosphomolybdate at 450 °C:

$$2 \; CH_2{=}CH{-}CH_3 + 2 \; NH_3 + 3 \; O_2 \rightarrow 2 \; CH_2{=}CN{-}CN + 6 \; H_2O$$
propylene   acrylonitrile

It may seem odd that the methyl group is attacked while the double bond, usually so much more reactive, is unaffected. However, it is characteristic of hydrogen abstraction reactions of such compounds that hydrogen is preferentially removed from a carbon atom attached to an ethylenic carbon atom.

6 Colour and lustre

Without dyes, garments and household textiles would be dull as ditch-water—and much the same colour. Even in prehistoric times the craft of dyeing was wide-spread, and such colours as woad, indigo, Tyrian purple, and alizarin were obtained by extracting plants, lichens, and shellfish.

The coloured compounds used as dyestuffs have chemical structures such that they absorb visible light of certain wave-lengths and so transmit only selected wave-lengths to the eye. Not all such coloured compounds are dyestuffs, for a dyestuff must also possess a structure such that it can enter the fibre and thereafter remain within the fibre during use. Different fibres, as the following experiment shows, require different dye structures.

EXPERIMENT 14. Dyeing three colours from one bath.

Make up a dye bath containing 100 cm^3 water, 10 cm^3 0.9% formic acid, 0.25 g Solway Blue BN, 0.25 g Durazol Scarlet 2G, and 0.25 g Dispersol Fast Yellow G. Place fabrics of cotton (1–1.5 g), wool (2–3 g), and acetate or Terylene (1–1.5 g) in the bath, raise the temperature to the boil over $\frac{1}{2}$ hour and dye at the boil for 1 hour. Remove the fabrics and rinse under the tap. The cotton is red (direct dye), the wool dark blue (acid wool dye), and the acetate or Terylene yellow (disperse dye). Alternatively the yellow dye may be omitted and Terylene rather than acetate used; the fabrics then emerge red, white, and blue.

Early dyes were all natural. *Vat dyes* can be reduced to a soluble form, applied to the fibre, then allowed to oxidize back to the insoluble colour. *Mordant dyes* are soluble but form insoluble complexes with the hydroxides of metals such as aluminium or chromium, deposited within the fibre in a previous treatment. The soluble forms are weakly acidic, so they can migrate into the hydrophilic fibres like wool, silk, and cotton. Once oxidized or complexed with a metal they become so insoluble that cleaning processes fail to remove them. These two classes of dye are still important, but nowadays the dyes used are synthetic.

The first synthetic dyes depended on a different principle: they contained basic or acidic groups that interacted with the fibre itself. The protein of wool contains acidic groups, so *basic dyes* enter wool and form ionic bonds with the acidic groups strong enough to prevent removal of the dye during cleaning. The first useful synthetic dye, mauveine, discovered in 1856 by William Perkin, was a basic dye. Wool also contains basic amino groups, so *acid dyes* similarly form strong ionic links and are fast on wool. The protein of silk, like that of wool, contains both carboxylic acid and amino groups, with the result that silk too is dyed by both basic and acid dyes. Nylon-6 and -6,6 also contain amino groups as polymer end-groups, so they can be dyed with acid dyes.

EXPERIMENT 15. Effect of deamination on dyeing of wool.

To 50 cm³ of a cold 30% solution of acetic acid in water add 50 cm³ of a 40% solution of sodium nitrite in water. Leave one of two 5 g samples of undyed wool fabric or natural blonde hair in the solution for 2 days, then wash it thoroughly in distilled water. This treatment removes basic amino groups by the reaction

$$RNH_2 + HNO_2 \rightarrow ROH + N_2 + H_2O$$

Dissolve 0.3 g Solway Blue BN, 1 g Glauber's salt, and 0.15 cm³ glacial acetic acid in 500 cm³ water. Wet both samples thoroughly and place them in the dye bath. Heat to the boil during 30 min then boil for 30 min. Remove the samples and compare the colours.

Cellulose fibres such as cotton contain no carboxylic acid or amino groups, so they cannot form strong ionic bonds. They do contain a very large number of alcoholic hydroxyl and ether groups, which are able to form hydrogen bonds with suitable groups in a dyestuff. Each hydrogen bond is much weaker than the ionic links formed in acid or basic dyeing of the protein fibres, but formation of several hydrogen bonds between each dye molecule and the cellulose produces sufficient total bonding force to ensure retention of the dye within the fibre. The class known as *direct dyes* is particularly useful for cellulosic fibres. These are acid dyes, for they contain sulphonic acid groups that solubilize them in water, but generally they are of higher molecular weight and have a more extended structure than the acid wool dyes. Many are bisazo compounds like that illustrated; such structures give the maximum opportunity for hydrogen bond formation with the cellulose.

We saw when discussing wool that its disulphide cross-links could be broken by reducing agents and regenerated by oxidation. The same reversible reaction underlies the use of *sulphur dyes*. The chemical similarity to vat dyeing is also evident, for the dye is dissolved in the reducing agent sodium sulphide, migrates into cellulosic fibres in the reduced form, then re-oxidizes to an insoluble form in air or when treated with weak hydrogen peroxide.

Vat, mordant, and sulphur dyes are all useful because they are able to

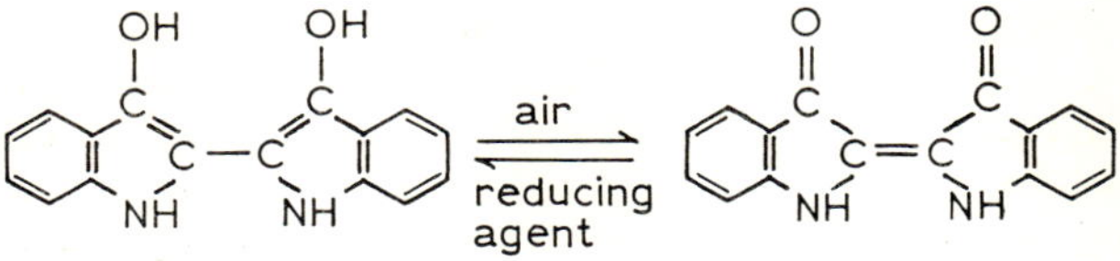

Colourless, alkali−
soluble leuco
compound

Indigotin, a vat dye
blue insoluble

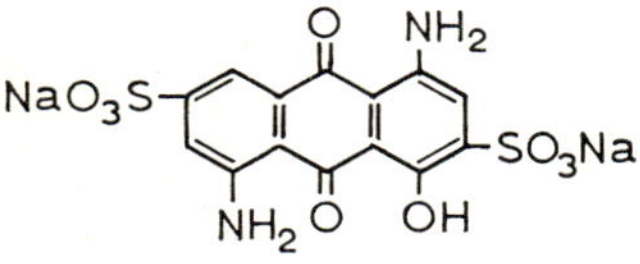

Solway Blue B, an acid dye

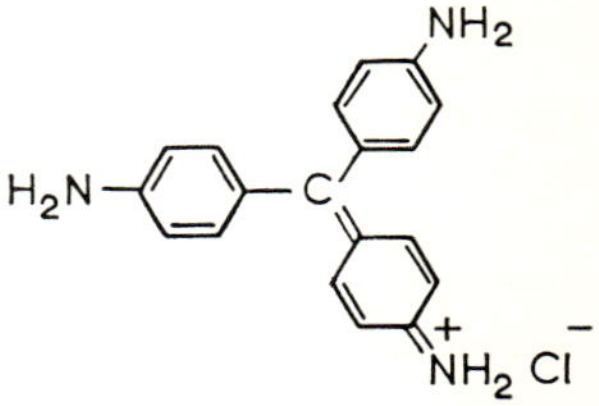

Pararosaniline, a basic dye

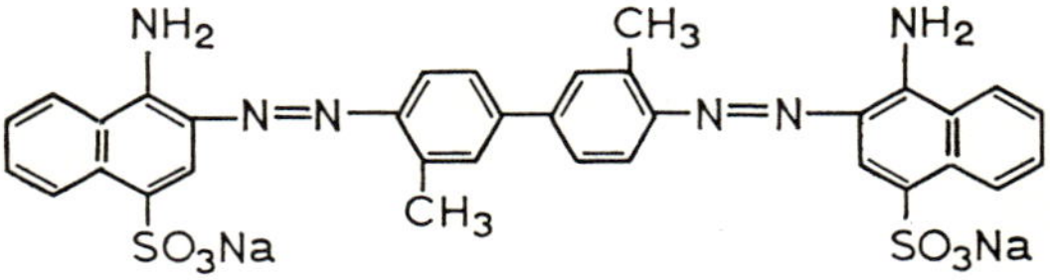

Benzopurpurin 4B, a direct dye

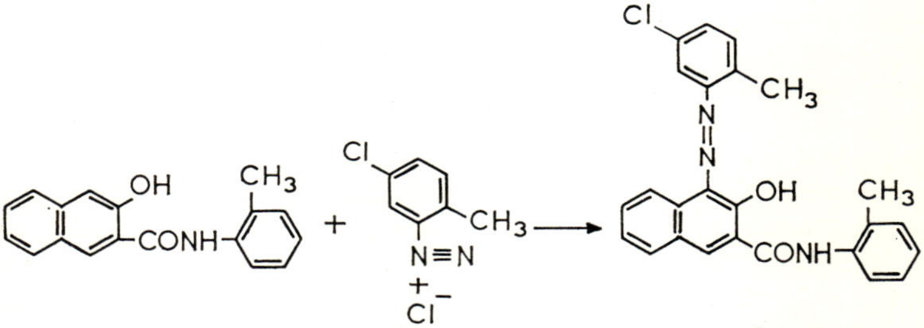

phenolic component   diazonium salt   an azoic dye

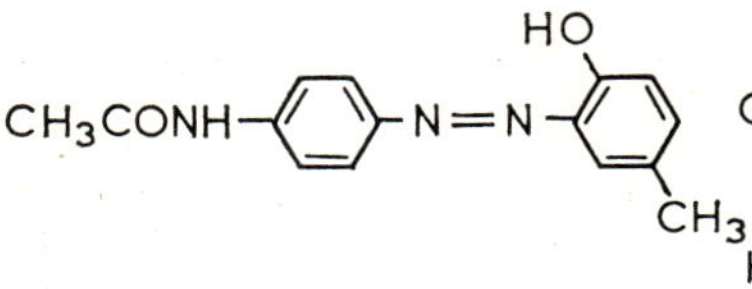

a disperse dye 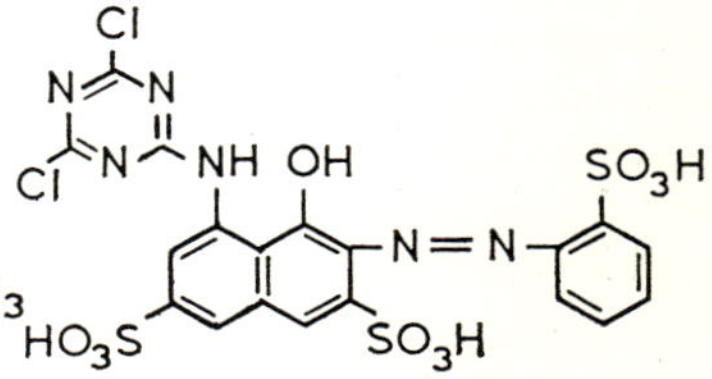   a. reactive dye

migrate into certain fibres in the soluble form but unable to migrate out once oxidized or complexed. Yet another way of achieving this result is to diffuse in separately two components that react together within the fibre to form the dye molecule. The *azoic dyes* are diazo compounds made by diffusing in first a phenolic compound then a diazonium salt; they react to form a dye which, being insoluble, has no tendency to migrate out.

A new class of dye was developed because of the difficulty of dyeing some of the hydrophobic man-made fibres, such as cellulose acetate, cellulose triacetate, and poly(ethylene terephthalate), with dyes that required ionizable groups or hydroxyl groups for their attachment. These *disperse dyes*, unlike all the previous classes in being only sparingly soluble in water, are applied from dispersions in water—hence their name. They are readily soluble in the fibre, very much more soluble than in water, so they diffuse into the fibre and remain there. The forces retaining them are hydrogen-bonding and polar interaction forces. Disperse dyes will not dye the very hydrophilic fibres such as cotton and wool, nor the very hydrophobic fibres such as polypropylene that lack all capacity for polar bond formation, but their use is not restricted to polyester and cellulose ester fibres. The nylons, for example, although readily acid-dyeable because of their basic end-groups, are more often dyed with disperse dyes.

A special problem is found with polyester fibres. Their saturation dye uptake with disperse dyes is greater than that of nylon-6,6 yet the rate of dyeing is very much less. Polyester has a high glass–rubber transition temperature (T_g) and the rate of diffusion of dye into the fibre is low at temperatures below T_g because chain mobility is very low. In order to dye at a useful rate it is necessary either to carry out the dyeing at a temperature above 100 °C—at a superatmospheric pressure in a closed vessel—or to add substances known as *carriers* to the dyebath. A typical carrier, such as orthophenyl phenol, is a swelling agent for the fibre. The rate of diffusion of dyestuff into the swollen fibre is so much higher than into the unswollen fibre that it becomes possible to dye at the boil and avoid the use of pressure.

One of the most effective ways of ensuring that a dye cannot be lost from a fibre is to attach it to the polymer by covalent bonds, which are much more durable than the ionic and weaker bonds used by most of the classes of dye so far considered. Since 1956 several types of *reactive dye* have been introduced, differing in the reactions used to attach them to the fibre. The Procion dyes were the first established. They dye cellulose fibres, and the principle of their operation will serve to illustrate the class. Alcohols, including the hydroxyl groups of cellulose, react with cyanuric chloride and its derivatives. If one, or even two, dye molecules are first attached to the cyanuric chloride, the remaining chloro-substituents will still react with cellulose. After reaction the dye molecules are firmly attached to the cellulose chain through a triazine ring:

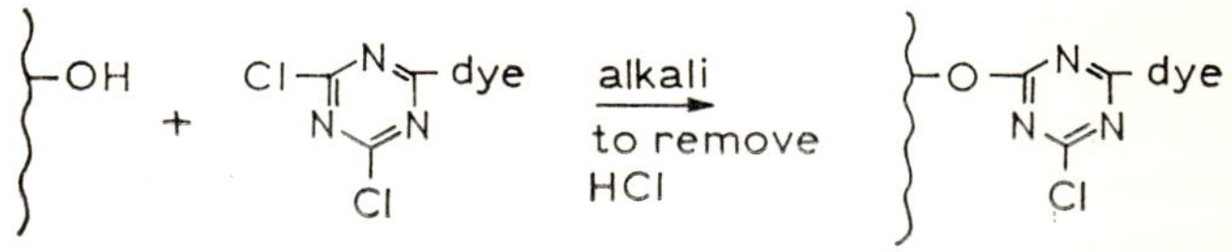

cellulose   reactive dye   dyed cellulose

Dyes have also been developed to react with the amino groups of protein and polyamide fibres. A variety of reactive groups have been devised to supplement those derived from cyanuric chloride, and yet more new ways of planting coloured molecules firmly in fibres are sure to be discovered.

Fibres with modified dyeability

So far we have related the types of dye with the chemical structures of the fibres as originally developed. But man-made fibres offer an extra possibility, since it is much easier to change their chemical structures to modify their dyeability than it is to change the behaviour of natural fibres. A simple example is modification of the polyester poly(ethylene terephthalate) to make it dye with basic dyes. The acidic groups required are provided by co-polymerizing into the polymer a small percentage of an ester containing sulphonate groups:

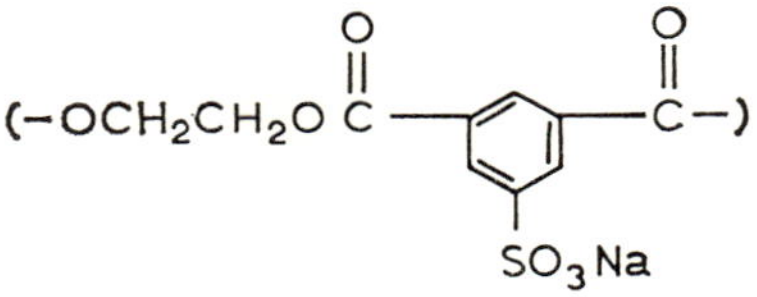

A more complicated example is modification of nylons. They can be made basic-dyeable by a modification analogous to that just described for polyester. They can also be made more acid-dyeable by introducing extra amino groups into the polymer chain, typically by replacing some hexamethylene diamine by N,N¹-bis-3-aminopropyl piperazine:

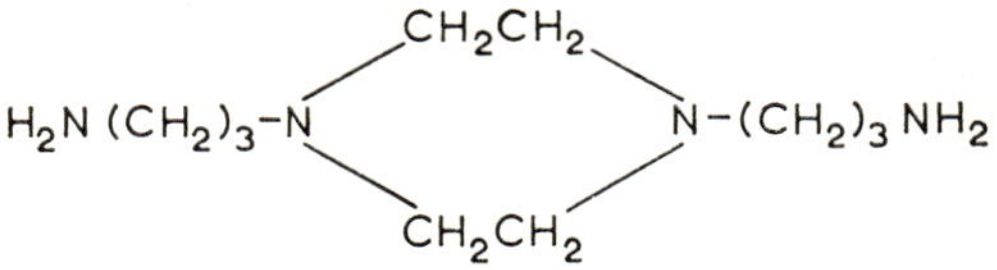

The most important example, however, is acrylic fibre. Unmodified polyacrylonitrile is only dyeable with a limited range of dyes using special techniques, yet commercial acrylic fibre is dyeable with disperse and either acid or basic dyes. This is achieved by copolymerization with a neutral co-monomer, such as methyl acrylate, and a basic or acidic co-monomer, such as a vinylpyridine or itaconic acid.

$$CH_2 = CH$$
$$|$$
$$COOCH_3$$

$$CH_2 = CH$$

$$CH_2 = C - COOH$$
$$|$$
$$CH_2 - COOH$$

methyl acrylate   4-vinylpyridine   itaconic acid

Yet another way of modifying the dyeing behaviour of a man-made fibre is to introduce an additive that is dispersed in the fibre and reacts with the desired class of dye. Some fibres, such as polypropylene, possess so little retentive power for all the usual classes of dye that incorporation of an additive is the only practical way of making them dyeable. Dispersing a nickel or aluminium salt in the fibre provides sites for mordant dyes; dispersing a polymer containing a high concentration of basic groups provides sites for acid dyes.

Pigmentation

One last, and very important, method of colouring fibres is available only for man-made fibres. Very insoluble, finely divided coloured compounds known as *pigments* are dispersed in the solution or melt of the polymer before it is spun into fibre. The pigment particles end up dispersed throughout the fibre. Because of their complete insolubility they are locked in so the colour is very fast.

Although a wide range of pigments is available, two kinds are much more important than the others. One is carbon black, partly because black and grey are required in greater volume than any single colour, partly because dyed blacks are rather expensive. The other, the most important of all, is titanium dioxide, which has played a vital part in the development of man-made fibres. The early man-made fibres were bright and lustrous, because they transmitted light freely. At first this lustre was unusual enough to help sell them, but eventually the time came when it proved a limitation and a matt appearance more akin to that of the natural fibres was desirable. The most successful method of dulling viscose, discovered in 1929, was addition of colourless particles of refractive index very different from that of the polymer. The most useful of these *delustrants* ever since has been titanium dioxide.

7 The future

In 1940 the first synthetic fibres were newly commercialized: 11 million lb were sold. The following 30 years have seen their consumption increase by 20–30% each year, or a thousand-fold in all, yet there is little sign of the rate of growth slackening. In the same period consumption of regenerated fibres rose only three-fold, and of natural fibres by less than 50%. Figure 7.1 shows how synthetic fibres overtook regenerated fibres in 1968, and how these man-made fibres together seem likely to overtake the natural fibres in the 1970's.

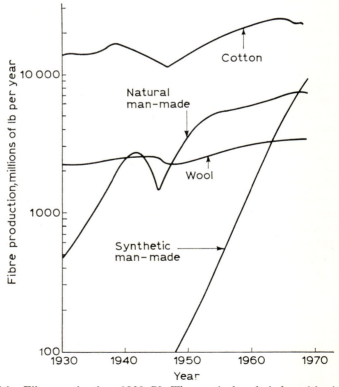

FIG. 7.1   Fibre production, 1930–70. The vertical scale is logarithmic.

Data from *Textile Organon*

Some of the reasons for the surge in synthetic fibre production have been reviewed: their strength, versatility, ease of care. Without such properties they could not have succeeded. But there have also been strong economic arguments in their favour. As production has increased, the advantages of large-scale production of the intermediates and polymers and the constant development of still better ways of making them have steadily reduced the prices of the fibres. The raw materials for the regenerated fibres, on the other hand, have not become cheaper, and natural fibre prices have fluctuated widely. Figure 7.2 shows how the price gap has narrowed.

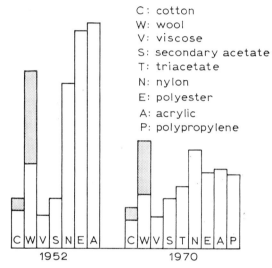

C: cotton
W: wool
V: viscose
S: secondary acetate
T: triacetate
N: nylon
E: polyester
A: acrylic
P: polypropylene

FIG. 7.2   Relative prices of textile staple fibres in 1952 and 1970. Shading shows range for natural fibres.

So far the advance in synthetics has not led to a reduction in natural fibre consumption. There is plenty of scope for synthetic fibre production to continue increasing at its present rate of nearly 1000 million lb. per year for many years without causing natural fibre consumption to fall. The developing nations of the world are building plants for synthetic fibres; these will not displace natural fibres, but will provide the increase in total fibre consumption associated with a rise in the standard of living. In the industrialized countries, the expansion will be based mainly on two features of synthetics: one, the variety which can be built in to them by the joint efforts of the chemist and the fibre technologist and which will lead, among other things, to new kinds of fibre assembly for use in traditional types of fabric; the other, the opportunity which synthetic fibre production provides to by-pass traditional, and often costly, methods of fabric manufacture.

Among the technical advances that will permit such developments are methods for producing fibres from two or more polymers combined side by side or as sheath and core in a single filament, thus:

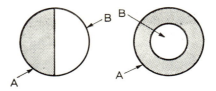

In the one case when the fibre is allowed to shrink the different shrinkages of the two components cause it to coil up giving a bulked yarn; in the other the fibre has the structural properties of the core with the surface properties of the sheath.

These developments will be accomplished in many cases by modifying the chemical structure of the fibres, but without requiring totally new polymers. Nevertheless, new fibres with novel chemical structures will also be introduced, just as they are being introduced at present. Two recent examples are a polyester derived from para-hydroxy-ethoxy benzoic acid, and a polyamide derived from bis-para-aminocyclohexyl methane and dodecanedioic acid.

$$(-OCH_2 CH_2O - \langle \bigcirc \rangle - CO-)_n \qquad (-NH - \langle \bigcirc \rangle - CH_2 - \langle \bigcirc \rangle - NH CO(CH_2)_{10}CO-)_n$$

A-tell   Qiana

Some of these new fibres will die; some will remain small volume, rather specialized products; one or two may join the ranks of major fibres, will reach an annual output measured in thousands of millions of pounds weight and an annual value in hundreds of millions of pounds sterling, and will then sustain a chemical industry providing hundreds of thousands of tons of intermediates and polymer every year.

Some well-known trade-names

Viscose rayon	Evlan, Sarille, Fibro, Vincel
Cuprammonium rayon	Cuprama, Cupresa
Secondary cellulose acetate	Acele, Dicel, Estron, Seraceta
Cellulose triacetate	Tricel, Arnel
Nylon-6,6	Bri-Nylon
Nylon-6	Celon, Perlon, Enkalon, Caprolan
Polyester	Terylene, Crimplene, Dacron, Trevira, Fortrel, Terlenka, Kodel
Polypropylene	Ulstron, Herculon, Meraklon, Tritor
Acrylic	Courtelle, Acrilan, Orlon, Dralon
Modacrylic	Teklan, Dynel
Spandex	Lycra, Vyrene, Spanzelle, Elura
Poly(vinyl chloride)	Rhovyl, Thermovyl, Fibravyl
Poly(vinyl alcohol)	Kuralon

Further reading

MONCRIEFF, R. W., *Man Made Fibres*, 5th Edn., Heywood, London (1970).

PETERS, R. H., *Textile Chemistry: Part I The Chemistry of Fibres*, Elsevier, Amsterdam (1963).

MARK, H. F., CERNIA, E., and ATLAS, S. M., eds., *Man-Made Fibers: Science and Technology*, Interscience, New York. (3 Vols., 1967–69.)

GOODMAN, I., *Synthetic Fibre-forming Polymers*, Royal Institute of Chemistry, London (1967).

CAROTHERS, W. H., and HILL, J. W., *J. Am. chem. Soc.* 54, 1559–87 (1932).

WARD, K., ed., *Chemisty and Chemical Technology of Cotton*, Interscience, New York (1955).

ALEXANDER, P., and HUDSON, R. F., *Wool, its Chemistry and Physics*, Chapman and Hall, London (1954).

VON BERGEN, W., ed., *Wool Handbook*, Vol. I, Interscience, New York. (3rd edn., 1963.)

KLARE, H., FRITZSCHE, E., and GRÖBE, V., *Synthetische Fasern aus Polyamiden*, Akademie-Verlag, Berlin (1963).

LUDEWIG, H., *Polyesterfasern*, Akademie-Verlag, Berlin (1965).

GOODMAN, I., and RHYS, J. A., *Polyesters, Vol. I. Saturated Polyesters*, Iliffe, London (1965).

SORENSON, W. R., and CAMPBELL, T. W., *Preparative Methods of Polymer Chemistry*, Interscience, New York. (2nd edn., 1968.)

MARK, H. F., GAYLORD, N. G., and BIKALES, N. M., eds., *Encyclopedia of Polymer Science and Technology*, Interscience, New York (1964 in progress), contains authoritative articles about general fibre topics and individual fibres. Figures 3.1, 3.2, and 3.4 are adapted from the article 'Man-made fibres: manufacture' in Volume 8.